"This book became an instant classic in the literature of [...] In this third edition, David Mann updates and expands [...] additional years of valuable experience and expertise derived from his very active, multi-industry consultancy. I have benefitted greatly from his writing and wholeheartedly recommend this book to be top-of-the desk of any serious Lean practitioner or performance transformation leader."

Raymond C. Floyd

President and CEO, Plasco Energy Group, recipient of the Shingo Prize for both Operational Excellence and Professional Publication; the only recipient of the Industry Week *"America's Best" designation in both process and discrete operations; and an Inductee to the Manufacturing Hall of Fame.*

"David Mann builds substantially on his seminal work on the Lean management system. The book is full of new insight and polishes the most important ideas about Lean management. The new chapter on engaging executive leadership alone is worth the price of the book."

Peter Ward

Professor & Richard M. Ross Chair in Management, Chair, Department of Management Sciences, Co-Director, Center for Operational Excellence, Fisher College of Business, The Ohio State University

"This book has long been my 'go-to' guide on Lean management practices that help create a culture of continuous improvement and excellence. I have recommended the book to countless healthcare leaders who rave about how helpful it is in translating Lean principles into daily management behaviors. The healthcare examples make it even more relevant as a must read for any hospital leader who aims to move beyond Lean tools..."

Mark Graban

Author of *Lean Hospitals*, co-author of *Healthcare Kaizen* and *The Executive Guide to Healthcare Kaizen*

"David has once again taken the topics that trip us up and put structure and guidance around them. His new work on executive involvement is worth the price of the book all by itself. Many of us have struggled with this topic and David provides a path to success."

Elizabeth M. King

Vice President Organizational Effectiveness, ESCO Corp

"As more companies outside the manufacturing sector pursue Lean transformations, *Creating a Lean Culture* is as critical a resource as ever. Breaking down silos and navigating tricky internecine politics remain a momentous challenge, and Mann's case-based insights are an invaluable tool."

Peg Pennington
Executive Director, Center for Operational Excellence,
Fisher College of Business, The Ohio State University

"My executives are not on board! This issue, familiar to Lean Champions everywhere, finds its resolution here. Mann provides the methodology and tools for successful executive engagement, ensuring their sustained commitment to the Lean Enterprise transformation."

Patrick O'Connell
Former Operations VP, AccuRounds

"In this third edition, Mann shares valuable new insights learned a decade after he first introduced the critical concept of manager standard work as a keystone for Lean transformation. Particularly important for top managers and change leaders are new sections on non-production application of manager standard work and keys to long-term engagement of top management.

Bruce Hamilton
President, GBMP

"In the ground-breaking first edition of this book, Mann offered the missing link to sustaining a Lean conversion, establishing himself as the father of Lean Management. In this latest edition, he builds on the principles and offers scores of real world examples that can be directly applied—it's like having your own personal sensei who coaches you on how to engage leadership for a sustainable Lean conversion, including what works and what to avoid."

Mark Edmondson
Lean Practitioner and Enterprise Change Agent

"David has an amazing ability to articulate the most important aspects of Lean that are often ignored by organizations—The Senior Leader involvement. An approach to create an effective, repeatable approach for engaging executives with Lean on their own terms will not be found in any other book or seminar in a manner, or with the result, that David's provides. Definitely a "must read" for leaders in organizations that are serious about their commitment to Lean."

Gary Olson
Director of Operational Excellence, JunoPacific Inc.
—A Cretex Medical Company

CREATING A LEAN CULTURE

Tools to Sustain Lean Conversions

THIRD EDITION

DAVID MANN

CRC Press
Taylor & Francis Group
Boca Raton   London   New York

CRC Press is an imprint of the
Taylor & Francis Group, an **informa** business

A PRODUCTIVITY PRESS BOOK

CRC Press
Taylor & Francis Group
6000 Broken Sound Parkway NW, Suite 300
Boca Raton, FL 33487-2742

© 2015 by Taylor & Francis Group, LLC
CRC Press is an imprint of Taylor & Francis Group, an Informa business

No claim to original U.S. Government works

Printed on acid-free paper
Version Date: 20140501

International Standard Book Number-13: 978-1-4822-4323-9 (Paperback)

Library of Congress Cataloging-in-Publication Data

Mann, David, 1947-
  Creating a lean culture : tools to sustain lean conversions / David Mann. -- Third edition.
    pages cm
  Includes bibliographical references and index.
  ISBN 978-1-4822-4323-9 (paperback)
  1. Organizational effectiveness. 2. Industrial efficiency. 3. Organizational change. 4. Corporate culture. I. Title.

HD58.9.M365 2015
658.4'01--dc23   2014015711

Visit the Taylor & Francis Web site at
http://www.taylorandfrancis.com

and the CRC Press Web site at
http://www.crcpress.com

To my wife, Jan, and our daughters, Kate and Elizabeth.

Without your love, support, encouragement, and especially patience, I could not have written this book.

Contents

SECTION II LEARNING LEAN MANAGEMENT AND PRODUCTION: SUPPORTING ELEMENTS OF LEAN MANAGEMENT

List of Figures

List of Tables

Acknowledgments

I owe much to many people with, for, and from whom I have worked and learned—bosses, clients, colleagues, sensei. I will try to acknowledge them here, and apologize to any I miss. Any shortcomings in this book are fully and exclusively my own responsibility.

The Association for Manufacturing Excellence (AME) gave me access to a wider professional audience than I could otherwise have had, and provided vehicles for reaching the larger manufacturing community.

Steelcase, Inc. gave me the opportunity to experiment with and develop many of the ideas and approaches to Lean management contained in these pages. At Steelcase, I had the good fortune to work for several executives who supported the idea that there had to be more to Lean production than the application of Henry Ford's and Toyota's industrial engineering techniques. Among them were Adolph Bessler, Rob Burch, and especially Mark Baker. I am grateful to all of them.

Mark Baker, Mark Berghoef, John Duba, and Scott McDuffee read early drafts of the first edition's manuscript and provided helpful critique and suggestions. I appreciate their willingness to take on this task and their thoughtful comments.

I had terrific internal clients to work with at Steelcase while developing (actually, co-developing) and testing the concepts and tools of Lean management. Among them were Shanda Bedoian, Mark Berghoef, Dave Greene, Dave McLenithan, John Mancuso, Kevin Meagher, Didier Rabino, and Jane Velthouse.

I learned a lot about Lean thinking from sensei Tom Luyster and Marek Piatkowski and am thankful they shared their knowledge and perspectives with me.

My colleagues on our Lean journey have been important in helping develop and refine much of what is in this book. Scott McDuffee, Pat Nally, Bob O'Neill, and Dave Rottiers all contributed to what has emerged as Lean management. John Duba and Ken Knister have been constant sounding boards and valued collaborators over the years.

Maura May of Productivity Press took a chance on the basis of a presentation at the AME conference in 2003 and provided welcome and needed encouragement throughout the writing process of the first edition of this book. Ruth Mills edited the manuscript and made refinements that make the book more accessible to readers.

Nancy Hickey, one of the senior executive steering team for the Lean initiative I led at Steelcase, provided me with clear advice and valuable insight that opened new thinking on engaging executives in Lean by actively involving them in it.

I have been fortunate since this book was first published to work with and visit many organizations. I only hope they have learned half as much from me as I have learned from them.

Finally, I want to acknowledge and thank my first, primary, and continuing editor, my spouse, Jan, who taught me that everybody needs an editor, but probably does not recognize that few ever get one as gifted as she is.

Introduction

Creating a Lean Culture addresses a gap in the literature on Lean production and the Toyota Production System (TPS). This book introduces a Lean system for management, first in concept and then in concrete detail. Lean management is a crucial ingredient for successful Lean conversions. Yet, the standard books on Lean either don't cover it or only hint at implementing new ways to manage in a Lean environment. *Creating a Lean Culture* provides the rationale and then a practical guide to implementing the missing link you'll need to sustain your Lean implementation—a Lean management system.

Lean production, based on the much admired Toyota Production System, has proved to be an unbeatable way to organize production operations. As I and others have learned since the first edition of this book was published, "production operations" exist well beyond the world of discrete manufacturing. You could fairly say anything of value, whether physical goods or services, is the outcome of a process that produced it. So, Lean approaches also work to improve process industries, technical-professional work, healthcare services, and a range of transactional service industries.

The key concepts of Lean are easily grasped, and relative to most technical engineering projects, Lean designs are easily implemented. Yet the majority of attempts to implement Lean production end in disappointing outcomes, and declarations like "Lean won't work here," or "with our people," or "in our industry," or "with our product/process," and so on.

Why, when it seems so simple, are successful Lean implementations so difficult to achieve? The answer is in an overlooked but crucial aspect of Lean. It requires an almost completely different approach in *day-to-day and hour-to-hour management*, compared to anything with which leaders in conventional environments are familiar or comfortable.

I'm a social scientist, an organizational psychologist. By conventional measure, I've been in the "wrong" place for someone with my training and background for the past 25 years. It has turned out to be absolutely the right place for me to be immersed in production operations and directly involved in supporting more than 40 brownfield Lean conversion projects, large and small, and over 100 business process value stream improvement projects. Through much trial and error in these experiences, I've come to recognize a common, but typically overlooked, element in batch-to-Lean or conventional-to-Lean conversions. This element is a Lean management system. Not only is it necessary to sustain new Lean conversions, but it also accounts for the differences between failed and successful implementations.

This book lays out the components of Lean management, how they work together, and how to implement the process. *Creating a Lean Culture* maps a course for leaders implementing Lean management to guide them through the cultural minefields in batch-to-Lean conversions.

Lessons Learned since the Second Edition

I have been fortunate in the five years since the second edition published in 2010 to have worked, visited, and corresponded with many people and organizations that have taken ideas from the book and run with them. An informal community of practice has developed around the recognition that to be sustained, Lean tools need the support of a new approach to management—a Lean management system. From the perspective of ten years since the original publication, it is safe to say that the ideas and practices of Lean management have gained traction within the Lean community and in organizations across fields, industries, and geographies.

I remain struck by the wide application of Lean production and Lean management across economies around the world, whether in software and pharmaceuticals development, process industries, healthcare delivery, public and private service industries, and manufacturing of every kind, scale, and technology. The approaches people have taken to Lean management turn out to be quite imaginative.

One conclusion I have come to is that there are many "right" or effective ways to do Lean management. Another conclusion is that there are some things to avoid. In any endeavor, negative feedback is often the most valuable. So, some of what is new in this edition are new or reemphasized lessons from the experiences of Lean implementers with whom I have met and worked in the past five years. Some are things to avoid that

I have encountered in these Lean management applications. Other new content has come from seeing problems with even well-conceived and implemented Lean management systems. This new content particularly reflects widely experienced difficulty in engaging executives with Lean implementation and leading the Lean initiative in ways that go beyond initial rhetorical support.

The next section briefly describes the new material in this edition. Just as "lowering the water" really *does* expose the next set of rocks, these lessons have brought home again for me the reality of Lean as a journey. At our Lean best, we are all students in a boat on the same river together; the rocks are not always obvious to us, and this new edition attempts to provide a much needed "map" for the journey.

What's New in the Third Edition

■ The most substantial addition to this third edition is a completely new chapter, Chapter 8, on engaging executives. Chapter 8 suggests most executives don't want to learn Lean to the depth and in the manner Lean practitioners learned Lean. Instead, the chapter presents a Lean management-based method, process, and tools for involving executives in leading and sustaining their organizations' Lean initiatives. Lack of executive engagement continues to be a troubling problem frequently encountered by internal Lean practitioners. This new chapter addresses this vexing problem with an approach developed from and tested by experience and illustrated with three new case studies.

■ Twenty-one new case studies throughout the book expand on existing as well as new concepts and settings, particularly in healthcare, administrative, and process industries, and involving executives in a Lean initiative.

■ Appendices A and B—the Lean management standards—include updated references to Lean management in process industries, administrative applications, and healthcare.

■ Appendix C, new with this edition, is a revised version of the Lean management standards, formatted for use in teaching and learning gemba walks, particularly with executives. All of these tools are available for free download in PDF format at www.dmannlean.com

■ And, you'll find study questions at the end of each chapter to aid in your own understanding of the material and for use in your organization's internal discussions.

■ Chapter 2 includes a clearer description of how Lean management acts as a driver of continuous improvement beyond sustaining the initial gains from Lean production. The revised chapter, supported by new case illustrations, adds emphasis to advice on the sequence of implementing Lean management relative to Lean production, addressing the universal (though often unspoken) question: "What's in it for me?" The chapter addresses with greater clarity (and new case illustrations) how Lean management applies in process industries.

■ Chapter 3 includes a strengthened and case-illustrated recommendation on why it is rarely a good idea to begin Lean management with leader standard work—along with cases illustrating exceptions to the recommended standard sequence.

■ Chapter 4 adds healthcare examples, a case and visual control from a variable duration cardiology procedure area, and a visual control for inpatient discharge planning. Two cases from industry illustrate the importance of involving floor employees to supplement information collected from programmable logic controllers (PLCs) on automated lines, and avoiding reason codes used in place of operators' firsthand descriptions on production tracking and downtime charts when problems occur.

■ Chapter 5 adds a case illustration of accountability boards used in heavy maintenance turnaround activities in two continuous process industries. The chapter also addresses the often misplaced fascination with substituting flat-panel monitors to display accountability assignments and tracking charts.

■ In Chapter 6, Lean and Lean management applications in end-to-end office or enterprise business process value streams get additional focus on project structure, sequence, steps, and measures. Handling the internal political issues inherent in cross-functional value stream improvement projects also gets increased attention. Case studies in Chapter 6 describe the application of controlled release of work to minimize interruptions in a billing services department, and an application of value stream thinking to bring improvement to an entire business process in place of a failed piecemeal department-by-department approach to process improvement.

■ Chapter 7 makes a distinction between Lean management applied at tactical (or line management) levels compared with applications at strategic (or executive management) levels, illustrated with a case study that anticipates Chapter 8's treatment of involving senior executives with Lean in a way that is meaningful and valuable to them in their own terms.

■ Chapter 10 includes a new case study describing the application of a Lean production principle (separate in-cycle from out-cycle work) in a technical-professional group, illustrating the value of improving processes before (and instead of) changing organization structure.

■ Chapter 11 adds reference to semiautonomous "quick win" or "quick kill" improvement processes for smaller work groups.

■ Chapter 12 continues to remind that *you* are the most critical element in sustaining Lean and Lean management.

How This Book Is Organized

The book is organized into two sections. Part I answers the question: What is the Lean management system? This section details the principal elements of Lean management and includes Chapters 1 through 6, as summarized in Table I.1.

Table I.1   Principal Elements of Lean Management

Element	Key Characteristics	Chapters
Leader standard work	Daily checklists for line production leaders—team leaders, supervisors, and value stream managers—that state explicit expectations for what it means to focus on the process	3
Visual controls	Tracking charts and other visual tools that reflect actual performance compared with expected performance of virtually any process in a Lean operation—production and nonproduction alike	4, 6
Daily accountability process	Brief, structured, tiered meetings focused on performance with visual action assignments and follow-up to close gaps between actual results and expected performance	5
Discipline	Leaders themselves consistently following and following up on others' adherence to the processes that define the first three elements	1–6

Part II consists of Chapters 7 through 12 and covers how to learn Lean production and, especially, how to learn Lean management. The approaches to each are nearly identical. Nine attributes are described for leading a Lean conversion project and the important role played by these attributes in slightly—but importantly—different forms when leading an ongoing Lean production area.

New in this edition, Chapter 8 distinguishes between learning Lean (production and management) as an apprentice Lean practitioner and mastering Lean *management* as an executive. Most executives need not master Lean tools applications. Instead, mastery of Lean management equips executives to assess the health and fidelity of the Lean strategy through his or her chain of command. Part II also covers some supporting elements in the Lean management system, including the aspects of Lean management that focus on people-related issues. Table I.2 lists these and coordinates them with specific chapters. Finally, Part II focuses on what to do and how to sustain the Lean management and Lean production systems you have worked so hard (or will work so hard) to implement. In addition to guidance for leaders, this includes an audit of the Lean management system that can be readily adapted for use in your workplace.

Throughout the book are dozens of highlighted case studies, to show you how Lean management and Lean production work—or *should* work. The case studies are indexed by name, number, and page number in Table I.3.

And the book is chock full of sample visual ideas to jump-start your brainstorming for what might work best in your own organization.

Who Can Learn from This Book?

This book is for you if you are a leader at any level in an organization undergoing a Lean transformation. You could be an hourly leader of a production team, a supervisor of a department and team leaders, a value stream manager with supervisors and support group staff working in your value stream, or an executive in any area of responsibility—a plant or site or manufacturing general manager; an operations, division, or corporate executive, regardless of area of responsibility.

This book is also for you if you or your organization is contemplating a Lean transformation. You might not yet know much about Lean production. There are several clearly written and readily available books and articles you can turn to for an introduction. See the references section at the end of this book, and if you're unfamiliar with the terminology, scan the glossary, too.

Table I.2 Supporting Elements in the Lean Management System

Supporting Element	Key Characteristics	Chapter
Leadership tasks in an ongoing Lean operation	Subtle, but important differences between leading Lean conversion projects and leading ongoing Lean operations.	9
Executive involvement	Preparing executives to assess the health of Lean management instead of the tools of Lean production.	8
Learning Lean management	Work with a sensei; use the production area as the classroom through gemba walks.	7
Root cause analysis	Standard, basic tools to focus on eliminating the causes of problems rather than settling for workarounds that leave causes undisturbed.	10
Progressive discipline in a Lean environment	Applying discipline for performance as well as for conduct as a source of support for expected performance in a Lean environment.	11
Rapid response system	Procedures and technology for summoning quick help from support groups and management are important in finely balanced Lean operations; new relationships between support groups and production areas are an often overlooked critical factor for successful "911" systems.	10
Improvement process	How are improvement activities managed when they exceed the scope of the daily task assignment boards?	10
Appropriate automation	IT networks can be powerful tools in support of Lean production and Lean management; much of the power in IT for Lean consists of knowing when *not* to apply it.	9
Labor planning	A suite of four visual tools for planning the next day's work assignments, rotation plan, and unplanned absences.	11
HR policies	Changes to pay plans, expectations for rotation, applying discipline for problem performers, break schedules, communication processes, pay grades and classifications, and other "people" issues that help or hinder a Lean operation.	11
Assessing the status of Lean management	A five-level assessment on nine dimensions of Lean management (eight dimensions for office processes) to highlight areas needing attention to bring Lean management practices up to a self-sustaining level.	11, Appendices A, B, and C

Table I.3　Case Study Index

Case No.	Name	Area	Third Edition	Page No.
1.1	Backsliding in a Once Lean Emergency Room	Healthcare	X	6
1.2	Successful Implementation of Lean Management	Manufacturing		14
1.3	Grasping Reality—Healthcare	Healthcare	X	16
1.4	Grasping Reality—Admin	Admin	X	18
2.1	How Lean Management Fails When the Elements Don't Work Together	Manufacturing		34
2.2	Process Focus beyond Takt Time	Manufacturing		45
3.1	When Leader Standard Work Is Meaningful	Manufacturing		53
3.2	When Leader Standard Work Is Not Meaningful	Manufacturing	X	54
3.3	The Value Proposition for Leader Standard Work	Process industries	X	55
3.4	Leader Standard Work's Role in Creating and Maintaining Stability	Manufacturing		60
3.5	The Value of Sticking to the Plan	Manufacturing		61
3.6	Leader Standard Work—Compliance or Improvement?	Manufacturing	X	69
3.7	Improvement Is Everyone's Job	Manufacturing	X	70
4.1	Using Visual Controls to Improve Performance	Manufacturing		77
4.2	Problems Linger When You Track but Don't Act	Manufacturing		80
4.3	Taking Improvement to a New, More Focused Level	Manufacturing		81
4.4	Make Sure Your Employees Know the Charts Are Not Micromanaging Them	Manufacturing		86

Table I.3 (*Continued*) Case Study Index

Case No.	Name	Area	Third Edition	Page No.
7.3	Don't Use Consultants as a Pair of Hands	Healthcare	X	177
7.4	How Gemba Walks Can Reveal Opportunities for Improvement	Manufacturing		179
7.5	Learning Lean Means Experiential Learning	Executives	X	183
8.1	Why Gemba Walk the Front Line	Executives	X	195
8.2	Engage Execs by Active Involvement in Lean	Executives	X	197
8.3	Developing Executives' Eyes for Waste	Executives		205
8.4	A Burning Platform for Leader Standard Work	Executives/process industries	X	208
9.1	Taking Lean to the Next Level	Manufacturing		224
9.2	Taking Control of Improvements	Manufacturing		226
9.3	Results-Only Focus Corrupts Process Measures	Manufacturing		232
10.1	New Leaders Uncover the Cause of Old Problems	Manufacturing		242
10.2	Long-Term Systematic Cause Analysis of an Assembly Line Problem	Manufacturing		248
10.3	Working with a Response System (Number 1)	Manufacturing		253
10.4	Working with a Response System (Number 2)	Manufacturing		254
10.5	Experiment with Changing Process before Structure	Technical professional	X	255
11.1	What Happens When Ideas Are Neglected	Manufacturing		267
11.2	Training by the Book or by the Floor?	Manufacturing		274
12.1	Don't Solve Production Problems by Heroics	Manufacturing		286

You might be an engineer or other technical professional concerned about those pesky humans who are likely to trip up your well-designed system. You will find constructive ways of dealing with these issues here, too. Or, you might be a support group manager with questions about the demands Lean production will make on you and your group.

This book is also for you if you are wondering what the heck happened to the last project you worked on, the Lean project that seemed to hold such promise, but just did not turn out as you hoped it would.

Perhaps least likely, you might be a social scientist interested in how manufacturing organizations undergo change and the conditions that either support or impede it.

Whoever you are, welcome and let's go!

WHAT IS THE LEAN MANAGEMENT SYSTEM? PRINCIPAL ELEMENTS OF LEAN MANAGEMENT

1

Chapter 1

The Missing Link in Lean: The Management System

Most prescriptions for *Lean production* are missing a critical ingredient: a Lean management system to sustain it. (By production, I mean any value-adding industrial, technical-professional, administrative, or healthcare process producing goods or services.) Lean management practices are like many other aspects of Lean: easy to grasp, but difficult to execute consistently. This book spells out the distinction between an organization's *culture* and its management system; provides a framework to see the differences between Lean and batch or conventional cultures; and details the practices, tools, and thinking for establishing Lean management. A *Lean management system* sustains the gains from implementing Lean production. And, Lean management extends the gains from Lean production by putting directly into practice an original Lean adage: Lean is first about *finding* waste (Shingo, in Dillon, 1987). In these ways, Lean cultures grow from robust Lean management systems, and this chapter shows how that happens.

Developing a Lean Culture

What is culture? Is it real? Should it be among the targets in your Lean implementation? As a working definition, consider culture in a work organization to be the sum of people's habits related to how they get their work done. You will see several examples in the next few pages. Given that, culture must be "real," right? Well, yes and no.

In scientific language, culture is a *hypothetical construct*. That is, culture is a label or idea—a concept we make up to organize and get a handle on what we have seen or experienced. It has been said that something is real if it has observable effects. Culture certainly fits the bill there. People talk about their company's culture all the time as a reason why they can or cannot do something. Keynote speakers refer to an organization's culture as enabling or inhibiting change or resistance. Annual reports proudly refer to company culture as an invaluable asset, and so on.

Should a company target its culture in its efforts to transform its production processes and all the positions—high and low—associated with it? It is tempting to answer yes, but that would be a mistake.

Culture is no more likely a target than the air we breathe. It is not something to target for change. Culture is an idea arising from experience. That is, our idea of the culture of a place or organization is a result of what we experience there. In this way, a company's culture is a *result* of its management system. The premise of this book is that culture is critical, and to change it, you have to change your management system.

So, focus on your management system, on targets you can see, such as leaders' behavior, specific expectations, tools, and routine practices. Lean production systems make this easier, because they emphasize explicitly defined processes and use visual controls.

Don't Wait—Start Now!

Do not wait for the "real work," the physical, procedural, or other changes of a Lean implementation, to be done before turning your attention to implementing the management system. Your Lean implementation depends on it to survive! Think of it this way: in successful brownfield conventional or mass-to-Lean conversions, no more than 20 percent of the effort involves the typical "what you see is what you get" physical changes. You install new layouts to establish flow, develop procedures to improve handoffs between departments, begin pull signaling, develop ways to pace production, and so on. An advanced version of an initial implementation would also include features such as visual methods to track production and capture misses or problems, start-up meetings, and standardized work posted at workstations. But all that only gets you to the 20 percent level, at most, and the likelihood of disappointment with the staying power of the changes you have implemented and disappointment in the results the new system produces. You might come to the conclusion that Lean production does not really fit with your business model, culture, or industry, or some similar explanation.

Well, of course a Lean implementation that's only 20 percent complete is not going to be so hot. You have only done the easiest one-fifth of the process! The remaining 80 percent of the required time and effort is made up of tasks that are less obvious and much more demanding. After the design/implementation project team finishes and moves on, a very different, more subtle sort of rearrangement remains to be done. As a leader, many things change for you: the information you need to rely on, your deeply ingrained work habits, your day-to-day and hour-to-hour routines, and the way you think about problems, managing work, and productivity. All of these and more have to be transformed for your Lean implementation to be a long-term success.

Your Personal Lean Journey

You have probably heard over and over that Lean is a journey. It is true, but the journey truly begins in earnest *after* the production floor has been rearranged, or procedures redefined. Most of the journey is internal, a mental recalibration and adjustment to a Lean world. On this journey you learn to impose on yourself the same kind of disciplined adherence to process you now expect of operators in following their standardized work or standardized procedures. As you continue on the journey you learn to focus with near obsessive intensity on the processes in your system. You learn to trust that results will take care of themselves when you take care of the process.

Without this internal work, the most typical outcome of Lean implementations is to reinforce old habits and ways of thinking. As with any new system, when the Lean process is turned on, a variety of problems suddenly appear. Without a Lean management system in place to support the new physical or procedural arrangements, people are left to rely on their old tricks for fooling the system, using familiar workarounds to get themselves out of trouble. This is as true for leaders as it is for operators. It is a path that leads swiftly away from a successful Lean conversion. Worse, once you have realized your mistake, it is an uphill battle to convince people to try again, that you are serious this time and will stick with the change. Most often, the result is merely a different layout or just another new flowchart. The promising Lean system becomes one more sad entry in the roster of failed change projects (Case Study 1.1).

The ER improvement project ended without implementing new and different practices, processes, tools, and behaviors on the part of all involved, from senior leaders to ER floor nurses and support staff members. The Lean tools implemented as part of the project were left to depend on the committed people who had served on the project team. The project's success was

CASE STUDY 1.1: BACKSLIDING IN A ONCE LEAN EMERGENCY ROOM

A downtown hospital in a mid-sized city was having problems in its emergency department—its ER. Patients often waited four or more hours to be seen. The ER was frequently so backed up it had to tell incoming ambulances to divert to other hospitals, this in a competitive environment where the majority of new patients, and the revenue they represented, were admitted to the hospital through the ER. Members of the hospital's board of directors included CEOs of local manufacturers that had experienced the benefits of converting conventional operations to Lean production. These board members suggested engaging a Lean consulting firm to improve the processes, throughput, and patient satisfaction in the ER.

The ER Lean project, nearly two years in duration, was a success by almost any measure; the Lean tools had worked—though Lean management had not been implemented. Among the improvements:

- The triage process had been standardized and streamlined. Performed by skilled, clinically experienced nurses, triage took just a few minutes.
- Triage fed two flow paths for "treat and release" patients (about 80 percent of patients) or "diagnose and admit" patients (about 20 percent of patients).
- Standardized procedures for room changeover between patients and for prepping patients for examination eliminated rework and delays, increasing throughput velocity.
- ER nurses were cross-trained to draw blood for lab tests, eliminating the wait for phlebotomists to do the draws.
- An x-ray machine was relocated to the ER, eliminating waiting for patient transport staff to arrive and the waste of transport itself.

The total effect eliminated waiting altogether. Patients in both flow paths were triaged and brought back to an ER bed immediately. The time from door to discharge for patients in the treat-and-release flow path had been brought down to just over two hours.

I toured the ER with a member of the hospital's internal continuous improvement team two years after the project. My guide had supported the project and consultants. Not much was left of the improvements.

- The ER charge nurse who had been the internal project team leader and champion had moved to a position elsewhere in the hospital.
- Other ER nurses had also moved, so now triage was handled by whichever nurses were available and took 15 minutes or longer.
- Nurses new to the ER were not trained for blood draws that met the lab's quality requirements, so phlebotomists were again being called to do the draws. This meant waiting for phlebotomists was again a regular occurrence.
- The imaging department had shifted the x-ray techs to duties in the central x-ray area, distant from the ER. This meant waiting for x-rays had returned, as did waiting for patient transport at both ends of the round-trip from ER to x-ray and back again.
- And, after the project's successful completion, the attention of the managers and executives who had been involved shifted to other, current priorities. The ER was left on its own and gradually returned to "the way we've always done things around here."

The ER's throughput was trending down, and patient wait time and door-to-discharge times were trending up. My guide and I agreed: the Lean tools had worked as always, but without a Lean management implementation, the tools had atrophied and their gains had faded.

virtually completely person dependent. The project concluded before the new processes, roles, standards, procedures, and expectations were documented. No ongoing monitoring process was put in place.

The Lean tools were left to survive on their own with no Lean management system for support. When the ER project team members moved on to different positions in the hospital or elsewhere, there was nothing left behind to provide direction, standards, expectations, and monitoring—the elements that any new process needs to survive, continue, and improve. As is usually the case, even the ER's objectively improved performance could not survive the reemergence of ingrained habits and the lack of attention by busy hospital executives.

Why is it that so many attempts to convert to Lean end in retreat and disappointment? It is a paradox: so many Lean implementations fail because Lean is too easy! That is, it is too easy to implement the physical or procedural trappings of Lean production while failing completely to notice the need for a parallel implementation of Lean management. It is too easy just to keep on managing the way we always have. Instead, for the new physical production process to be a success, managers must change from the habitual focus on results to a quite different and less obvious focus on *process* and all it entails.

Lean Management Focuses on Process

The Lean management system consists of the discipline, daily practices, and tools you need to establish and maintain a persistent, intensive focus on process. It is *the process focus* that sustains and extends Lean implementations. And the practices of Lean management create process focus. Here is how this happens:

1. The visual controls called for in a Lean management implementation represent Lean's emphasis on process. Timely maintenance of visuals provides physical evidence of leaders' discipline. Equally important, visuals are designed to capture abnormal process performance events, including information on misses, defects, interruptions, system failures, and abnormalities.
2. The visuals with their process data are brought to a standard, often daily accountability meeting. At the accountability session, misses are reviewed and, where appropriate, converted into task assignments. The assignments may call for cause analysis, implementing a temporary countermeasure to protect the process from failure, or putting in place a root cause solution.
3. Follow-ups on the actions resulting from accountability assignments become items added, at least on a temporary basis, to leaders' standard work. Is the countermeasure working? Has the root cause fix remained in place? Adding these items to leader standard work supports the integrity of the change made in response to the miss or failure first noted on the visual process control.

Lean Management and Continuous Improvement

The information on the visual control, when brought to the accountability meeting, results in the process change. The follow-up through leader standard work provides the mechanism to systematically sustain the change. This prevents the process change from fading and disappearing, which happens so often when follow-through is lacking.

If this sounds circular, it is designed to be. This virtuous circle starts with recording on visual tracking charts occurrences of delays, defects, and interruptions. This information leads to accountability task assignments for cause analysis and process improvements. Follow-up on these process changes—the improvements—is incorporated into leader standard work. Taken together, these three elements, visuals, accountability task assignments, and leader standard work, make a closed-loop system that creates process focus and results in process improvement (Figure 1.1).

Put more plainly, the virtuous circle from a Lean management implementation does not just sustain the gains from Lean production. In addition, it produces one of the most difficult to achieve aspects of Lean: continuous improvement. No separate program and no special incentives are needed beyond maintaining a focus on process *as it cycles*. It is an example of the adage: *take care of your process and the process will take care of you.*

As Lean management, with its closed-loop focus on process, becomes habitual, little by little—almost unnoticeably at first—a Lean culture begins to grow. The new Lean culture emerges as leaders replace the mindset to work around problems today, ignore their causes, and let tomorrow take care of itself—a mindset learned in our careers in conventional processes and organizations.

Let us step back for a moment to provide some context for the conversion to Lean production and the differences in management system between conventional or batch and Lean. Lean manufacturing is an idea whose time has come. Manufacturers, and now organizations in every sector the world over, have recognized the advantages in lead time, productivity, quality, and cost enjoyed by Lean competitors in industry after industry. Lean is spreading

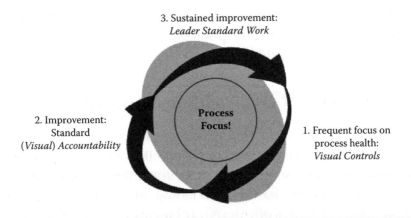

Figure 1.1 Lean management as a closed loop system creates process focus and drives improvement.

rapidly beyond the manufacturing sector, with applications in office process of all descriptions, healthcare, the private and public service sectors, product development, and IT.

One of the attractive features of Lean is that it is so easy to understand. Customer focus, value stream organization, standardized work, flow, pull, and continuous improvement are readily grasped. Second, Lean is typically not capital-intensive. In fact, Lean adherents prefer simple, single-purpose equipment with minimal automation. Lean scheduling systems are equally simple and inexpensive, rarely requiring much, if anything, in the way of incremental IT investment. Here, too, Leansters typically say less is more. In office, technical, and professional processes where the product is information, increasingly Lean improvements are turned to before making IT changes, with value stream maps used to define and document requirements for IT system changes. Finally, Lean layouts and material flows are relatively straightforward to design and implement, whether through redesign of entire value streams or more narrowly focused kaizen events.

Parallel Implementations of Lean Production and Lean Management

So, Lean production confers many advantages. It is easily grasped, requires minimal capital for equipment and IT support, and is relatively straightforward to implement. Yet, as I have just suggested, the experience of many, indeed most, companies that have attempted to convert to Lean production has been failure and retreat. Again, this is as true in office processes as in manufacturing. It seems so easy, yet success is so difficult! What is it about Lean that makes successful implementation so rare as to be newsworthy? Something, some crucial ingredient, must be missing from the standard list of steps in Lean conversions. The missing link is this: a parallel Lean implementation effort to convert management systems from conventional production to Lean.

The physical changes in a Lean conversion are easy to see: equipment gets moved, inventory is reduced and redeployed, and notable changes occur in material supply, production scheduling, and standardized methods. In an office, changes take place to processes, procedures, handoffs between departments, sequence of work steps, and measurements. The change in management systems is not as obvious. An orienting question about the Lean management system might be: Change *from* what *to* what?

Changing from Conventional Production

Think about management in a conventional batch-and-queue production operation. First and foremost, the focus is on results, on hitting the numbers:

- Did we meet the schedule for this day or this week?
- Did we discharge the patients on time?
- How many defective units were caught by quality inspections?
- Did we meet the target for customer response time?
- Did we hit our targets for material cost and production labor?

Managers in conventional systems track key indicators like these through monitoring and analyzing reports that summarize the previous period's (day, week, month, or quarter) data.

Managers attend many meetings to review production status and trouble-shoot problems. These meetings typically revolve around computer-generated reports that line managers and support group specialists pore over in conference rooms. Disagreements are common about which departments' reports to believe. In fact, it is not unusual to spend substantial time in meetings like these arguing about whose report is accurate. ("Your report says the material got here; my report says I never received it.") Sometimes you actually have to go to the production floor to look for or physically count what the computer says is supposed to be there—an extreme measure in an IT world.

The focus is usually retrospective, looking at what happened last reporting period, determining who or what messed up, and deciding how to recover. With more sophisticated IT systems, these data are accessible more or less in real time. Looking at a computer monitor, managers can see a numerical or even graphic representation of what the IT system presents as the precise state of their production or service process. This seems like an improvement, and often can be—if the data are accurate (not just precise), and you can sift through all the available data to identify the critical numbers to watch, and you know which are the critical questions to ask of the database.

Getting Rid of the "Do Whatever It Takes" Approach

When problems arise that threaten schedule completion in batch-and-queue systems, the common practice is do whatever it takes to meet the schedule, whether in an office, technical-professional, or factory setting. Expedite internal parts, pressure suppliers, airfreight late materials, expand the size of the hold queue, put on more people, pressure the inspectors, park admitted patients in the hallway, reorder missing parts with a fudge factor to make sure you get the few good ones you need, work the office staff late, or bring people back on the weekend. Just meet the schedule, hit the target! Tomorrow or next week, it is a new day with a new schedule and new challenges. Things that went wrong yesterday are typically forgotten in the press to meet today's demands. After all, today's schedule must be met!

In fact, most managers—in offices, healthcare, and manufacturing alike—have learned how to be successful in this kind of system. They know the workarounds and tricks to ensure success in an uncertain environment where the bottom might fall out in one of several areas on any given day. The tricks of the trade include secret stashes of extra raw material, work in process (WIP), people, and even equipment to be called on in time of need, not to mention bringing in people from elsewhere internally or from temp agencies. Never mind that all this is costly in the long run. In the short run, results are what matter and the numbers do not lie—you either met the schedule, hit the target, or you did not.

Lean Processes Need Lean Management

In Lean systems, results certainly matter, but the approach to achieving them differs sharply from conventional management methods. The difference in a Lean management system is the addition of a focus on process, as well as a focus on results. The premise is this: start by designing a process to produce specific results. If you have done a good job of designing the process and you maintain it, you will get the specified results. In concept, this is simply a matter of maintaining production at takt time—the specified target rate. If you do, you meet demand or achieve your goal. As you make improvements in the process, you should expect improved results.

A critical point is to think about the Lean management system as an integral element of the Lean process. Here is why. If the process were a perfect system, it would always run as designed and always produce consistent results. A real-world system requires periodic maintenance and occasional intervention and repair to continue producing results. The more complex the system, especially the more automated it is, the more maintenance and repair it requires. It may not seem like this should be true, but it is. In manufacturing, a more reliable and flexible solution usually is to rely less on automation and more on people and simpler equipment.

Relying on people brings its own set of issues. People require all sorts of "maintenance" and attention. Left to their own devices, people are prone to introduce all kinds of mischief—variation in the system that can take things far afield from the original design. In a newly or recently converted Lean system, people's deeply ingrained habits from "the way we always did things here" will continually emerge at any opportunity, often for years. If anything, Lean production is more vulnerable to these effects than conventional batch production, because of the tight interdependence and reliance on precise execution in Lean designs. This is one of the paradoxes in Lean. Lean execution is process dependent. But, the process-dependent execution depends on people's adherence to the Lean process design. This is why discipline is such an important factor in Lean processes. Without a high degree of discipline in a Lean process, chaos ensues in short order. Chaos prevention is where a Lean management system comes in.

Focusing on the Process Produces Results

Process focus works like this: if you want a process to produce the results for which it was designed, you have to pay attention to it. One of the first rules of process focus in a Lean operating environment is to regularly see the process operating with your own eyes. In classic Lean language this is known as grasping reality. It applies as much with administrative or health-care processes as in manufacturing.

The closer your position is to the production floor, the more time you should spend watching the process, verifying execution consistent with design, and following up when you observe nonstandard or abnormal conditions. Frontline team leaders, especially in high-volume, rapidly cycling discrete processes, should spend virtually all of their time either training workers in the process, maintaining the pace of production, monitoring standardized work or procedures in the process, or improving them.

Another way of thinking about this, and another paradox in Lean management, is that Lean managers are so focused on results that they cannot afford to take their eyes off the process they rely on to produce their results. Looking at what happened yesterday is way too late to do anything about yesterday's results. On the other hand, looking at what happened last hour, last pitch, last takt cycle or case, gives the chance to recover from an abnormal or nonstandard condition. But, that is only true if trained eyes (like a team leader's) are there to see the abnormality; the pertinent processes are well defined, clearly documented, and operating in a stable environment; and resources are available to respond in *real* real time. That is, someone must be available *right there* to respond *right now*!

Further, this means focusing on the process as it operates from beginning to end, not only at the completed component, finished goods, completed encounter, or other end-of-process outcome. That is why Lean designs require so many team leaders to maintain the process, to spot problems in upstream intermediate or subprocess areas, and to respond right away to prevent or minimize missing takt at the outlet end of the process. An integral part of the Lean management system is having the appropriate number of team leaders on the floor to focus on the process. It requires a leap of faith not to scrimp on this crucial part of the system. Having enough leaders available to monitor the process, react to problems, and work toward root cause solutions is an investment that pays off in business results. But at first, and from a conventional perspective, team leaders just look like more overhead (Case Study 1.2).

CASE STUDY 1.2: SUCCESSFUL IMPLEMENTATION OF LEAN MANAGEMENT

The testimony of a former leader of a plant's Lean team underscores this point. He had been involved from the outset of a Lean transformation initiative in one of the "mother ship" plants located within sight of his organization's corporate headquarters. He provided the technical Lean vision for transforming a classic batch-and-queue production process with 13 schedule points into a flow-and-pull value stream with only a single schedule point. The design was elegant and effective, though not without controversy. It was a radical departure not only in the way production flowed, but also in the way it was scheduled and in expectations for leaders' and operators' performance. He ended up taking a job to lead operations in one of the company's divisions located in a distant part of the country.

He was recruited there to bring the Lean focus he had learned in his previous assignment, and he did so with the full support of his division executive. In a relatively short time, roughly 24 months after arriving on the scene and working with the staff he inherited, his team had pulled off three waves of changes in the factory:

■ First, they picked up, moved, and rearranged every product line's production operations.

- Second, they conducted kaizens on all of the rearranged operations to increase their level of leanness.
- Third, they transformed the scheduling process from one based on inventory transactions and multiple material requirements planning (MRP) schedule points to a manual, visually controlled heijunka process for all but the handcrafted 15 percent of the plant's output.

These were major changes in physical arrangements and technical systems. In the previous batch-and-queue system, the time from releasing a manufacturing order to the floor completing a product took 20 working days. In the new flow-and-pull system, this total through-put time has been reduced to two days. In addition to the changes in manufacturing, it took equally significant changes throughout the supporting processes—HR and finance, database and production control, and others—to make this happen.

Clearly, this operations leader knew Lean philosophy, how to implement it, how to teach it, and how to lead it. Yet he is emphatic that, as important as all the physical and technical changes have been, the operation did not experience much measurable benefit until the implementation turned to the management system. He described it this way:

The new layouts really enabled the management system. It wasn't until we began focusing on it (the management system) that we began to see big increases in productivity. That came from paying attention to the process, implementing hour-by-hour production tracking, defining standard work for team leaders and supervisors, and following up on accountability for action on flow interrupters and improvements. We've seen a 36 percent increase in annual sales dollars per full-time equivalent employee, once we had the management system in operation. Much of that is attributable to Lean management.

Note: This increase in productivity came during a steep industry downturn in which this division saw a 35 percent decline in sales.

Engaging Executives with Lean: A Different Approach

In the orthodox ideal Lean scenario, taking time to monitor or spot-check the production process applies all the way up the chain of command, though with decreasing frequency and duration. In this scenario, senior operations leaders (e.g., in manufacturing: plant manager, site manager, operations VP) in an organization committed to a Lean transformation learn to visit the front line regularly (weekly at plant level to quarterly for a multisite operations VP) or to gemba walk (described in Chapter 7).

The rationale for this top-to-bottom eyes-on involvement is straightforward. That is, watching a process on the floor as it cycles and talking with those who do the work provide much deeper insight into the progress of a Lean initiative than reviewing spreadsheets in a conference room (Case Studies 1.3 and 1.4).

Gemba walking the floor individually with their subordinates allows operations executives to assess directly how well their subordinates grasp the reality of their Lean operations. Being on the floor periodically, operations executives can see and hear for themselves the depth to which frontline workers and their immediate leaders understand Lean, or if they are just keeping their heads down and going through the motions.

But my experience during the past ten years with many organizations' Lean projects has frequently departed from the orthodox Lean ideal. Typically, executives I encounter from areas other than operations are completely uninterested in learning the nuts and bolts of Lean production. I suppose that should not be a surprise. But here is what has been a surprise: this

---

CASE STUDY 1.3:   GRASPING REALITY—HEALTHCARE

A hospital VP responsible for ancillary services, mainly diagnostic testing and rehabilitation services, had successfully pushed to switch to a centralized appointment scheduling system housed at the distant health system headquarters. Central scheduling had been in place for four months. Prior to the switch, support staff members in the hospital's several testing areas and rehab departments scheduled the appointments. The switch was justified on the basis of projected cost savings.

A value stream improvement project on the patient referral and scheduling process revealed a consistent pattern of multiple phone contacts between central scheduling and patients. "Phone tag" is a familiar and costly problem for healthcare providers, and a source of delay, frustration, and dissatisfaction for patients. In addition, central schedulers frequently failed to provide patients with critical information, such as advance preparations required for the tests. So, many patients arrived for their tests without having taken the required advance steps, such as fasting for 12 hours. When that happened, patients were told to go home. The test had to be rescheduled, with another round of phone tag.

After these repeated frustrations, more than a few patients were choosing to have the tests at a nearby competing hospital, eating into critically needed revenue.

The value stream team recommended going to an improved local scheduling process. Their plan included concrete proposals to eliminate waiting and duplicate efforts in the local process by taking advantage of the hospital's recently upgraded IT capabilities, and adapting internally demonstrated Lean best practices across several administrative processes.

In a project team workshop report out, the ancillary services VP strongly and forcefully objected, challenging the accuracy of the project team's information and the feasibility of its plan. In effect, she declared that there would be no changes to the central scheduling arrangement.

During this confrontation, the department director responsible for managing the relationship with central scheduling (who was a project team member and subordinate of the VP) said nothing. The next day, she met with the VP and walked her through a thick stack of patient complaints and emails between her and the central scheduling manager reflecting her repeated unsuccessful attempts to fix the problems, the dissatisfaction of patients, and the repeated extra work required of her staff. The VP, who only reviewed costs versus plan from central scheduling, had no idea of the problems, staff time, patient dissatisfaction, and accelerating trend of lost business. Those things were not on her spreadsheet.

To her credit, she quickly recognized the reality experienced by patients and her staff. She agreed to support a trial implementation of the project team's plan. The team's plan in practice produced better financial results and better staff and patient satisfaction than projected. As importantly, the VP learned a lesson about going to the place, talking with the people, and observing the process as it cycles. In short, she learned the importance of grasping reality.

CASE STUDY 1.4: GRASPING REALITY—ADMIN

In a related case, an IT and customer service director at an international regional manufacturing operation had been responsible for implementing a new enterprise resource planning (ERP) system to track and manage sales fulfillment, customs and shipping documentation, and customer service. Issues in the customer service process prompted him to request a value stream improvement project in this process. The project workshop began with creating a value stream map created by sales fulfillment, documentation, and customer service reps, the people who did the work in question.

As the map took shape, the director began arguing with his staff members, saying that the ERP software implemented six weeks earlier indicated the process worked differently than showed on the map. To prove his point, he brought out the flowcharts he used to design the ERP system. The frontline staff members patiently talked him through how they actually did the work, using examples from customer cases they handled earlier that same day to support what they had drawn on the map.

The director finally had to agree. He thought he knew how the work was done. He knew how he had designed the process and how he wanted it to work. But even among only three reps and only six weeks after the ERP system implementation, the work process had changed in response to the daily experience of the frontline staffers working with customers.

The director's abstract view did not match reality. To resolve the customer service problems that prompted him to sponsor the value stream project, he first had to grasp the reality of how the work was actually done.

lack of interest is often true of senior operations executives, many of whom have engineering or operating backgrounds, and who express what I believe is genuine support for their organizations' Lean strategy and Lean initiative.

Understanding the Executive Perspective

With time and reflection, however, I believe I have come to understand their perspective. Their responsibilities, like those of their peer execs in other areas, have grown far beyond day-to-day matters. Their calendars are jammed with obligations and demands from across the organization and from outside it as well. While many have technical backgrounds, experience with Lean was rarely part of it. Further, in their current positions, they rely on Lean implementers and others to worry about the tactical, technical nuts and bolts of Lean implementation, training, assessment, and coaching. The executive's day-to-day responsibilities involve strategy and strategic execution on a broad scale, not technical nuts and bolts.

So, I am no longer surprised to find senior operations and other executives politely aloof, disconnected from the Lean initiative. Likewise, it is not unusual to find internal Lean resources frustrated with the executive disconnect and with the Leansters' inability to engage them. For sustaining a Lean initiative, this frustration is a symptom of a serious problem that threatens continuation of the Lean initiative.

Without the active support of engaged executives, a Lean initiative is living on borrowed time, no matter how good the results it has produced. Without engagement, execs' attention tends to move on to the next new "big idea" in management, or the business cycle prompts belt tightening and Lean resources are seen as expendable overhead, or a "new sheriff" arrives on the scene—a new executive with a different emphasis. In any of these cases, the Lean initiative withers.

The root of the problem, I believe, lies in the orthodox Lean ideal, and specifically this aspect, namely, that everyone in the ideal Lean organization gets brought up through the ranks to become a Lean thinker and practitioner. Toyota and perhaps Danaher in the United States represent this ideal. Both have long-standing top-to-bottom emphasis on Lean thinking and application. At Danaher for over 30 years and at Toyota for over 50 years, leaders have been brought up as Lean thinkers and practitioners embedded in a Lean culture.

The rest of us are struggling to transform conventional leaders into Lean thinkers and an often deeply conventional culture into a Lean one. And, this transformation is supposed to happen in the midst of turbulent technological and regulatory change and an ever-shifting, globally competitive business environment.

Lean Orthodoxy—Not the Answer

Lean implementers—the Lean geeks and believers, including me—have largely been taught the Lean ideal. That is, the one true way to learn Lean is from the bottom up, by personal hands-on implementation of the

Lean tools yourself. Now, that may well be accurate, especially for those who will be implementing Lean and managing Lean operations day to day. The problem with this orthodoxy is that most of us Leansters approach the task of engaging executives in Lean by asking them to learn Lean the way we did, the "hard way" or from the bottom up.

In 1970 Walt Kelly, a cartoonist, famously wrote: "We have met the enemy and he is us!" Put another way, the first principle of Lean (Womack and Jones, 1996) is that value is defined from the point of view of the customer. Much of the difficulty Lean practitioners experience in trying to engage executives has simply been our failure to see execs as our customers, and understand Lean is new to them and they do not know what to do to support it. The executives we want to engage are our customers, but we have failed to recognize where they are starting from, and what they value.

Fortunately, Lean management offers a way out of this problem.

To engage executives, Lean management makes a key distinction. That is, the way most executives can sustain and lead a Lean initiative is different from the way line managers and Lean resources support Lean. This distinction is based on this critical proposition: when the Lean management system is healthy, the Lean production system will be healthy, and when Lean management is weak, Lean production will be faltering and in trouble.

The role in Lean for line leaders and Lean implementers is tactical: which tools to apply, how to keep the new process working, and how to improve it. The role for senior leaders is strategic: what should their strategy be (Lean, in this case), how to execute the strategy effectively, and how to redirect it when it stumbles. In the strategic role for executives in Lean, their task involves developing mastery of the Lean management system as the basis, when on the floor, for assessing the health of Lean management, and through it the health of the Lean production initiative.

Why does this make sense for executives and how does it connect with their executive managerial and strategic responsibilities?

If executives find weakness in Lean management's floor execution, it flags a weakness somewhere in *their organization's* execution of the Lean strategy. Say the frontline people tell the exec that production tracking is a joke; nobody does anything to address problems on the tracking charts. Where is the problem? With the frontline people or somewhere above them in the chain of command?

Tending to the Chain of Command

That flag should prompt the exec to diagnose which link or links in his or her chain of command need attention and to take the appropriate action, such as resetting expectations, floor-based coaching, or if all else fails, disciplinary action. This is a management responsibility unique to those in senior executive positions, one that engages them in the Lean initiative in a way that only they can perform and which directly connects with their managerial and strategic responsibilities.

Executives enthusiastically engage in Lean after experiencing the proposition that healthy Lean management means healthy Lean production. They learn to assess Lean management's health through a structured, repeatable, gemba-based, mastery-tested, hands-on process in one-on-one coaching by an appropriately prepared internal Lean resource. This approach and process successfully engages executives in a personally meaningful way, one that provides real value to them in their own terms. They are purely managers, and the issues they identify through assessing Lean management are important ones, theirs to address as managers. The effect is true for senior operations leaders as well as with their peers in other areas of responsibility.

Chapter 8 further details this rationale and describes the processes, roles, tasks, and tools for Lean resources to take executives through this process. In it, executives in any kind of organization master Lean management and assess and address gaps in the integrity of their Lean strategy's deployment down their chain of command. They learn how and what to do to lead and sustain the Lean initiative in a way that not only fits their positions but also resonates with them personally.

Measuring the Process against Expected Outcomes

Unlike managing in a results-focused system, process focus implies frequent measurement against expected intermediate outcomes. As necessary, interventions can be started before the end results are affected. A corollary of frequent measurement at multiple intermediate steps in a Lean process is that data are readily available to aid quick diagnosis of problems, spur immediate remedial action, and eventually eliminate root causes of problems. This is one aspect of continuous improvement. Rather than waiting for problems to develop, you are constantly monitoring for early signs of developing troubles. It is especially the case for small problems of the sort that go unnoticed in end-of-the-period reporting. And, you are primed to take quick action to identify and eliminate the causes of problems. This is an illustration of the adage that Lean is first about *finding* problems.

Contrast this with the conventional batch production culture in which most supervisors expect various unpredictable problems and have earned their spurs by being able to work around them to get out the day's schedule.

A new management system is called for in Lean conversions, because Lean processes are much more tightly interdependent than conventional systems. Lean processes are designed not to rely on the extras stashed away in conventional systems to bail things out in a pinch. Even so, things go wrong in Lean systems just as they do in mass systems. By design, a Lean process has little unaccounted for slack in the system to fall back on. Because of that, Lean processes require far more attention to disciplined cycle-by-cycle operation to ensure the process stays in a stable state. Otherwise, the process will fail to hit its goals and fail to deliver the business results so important in any kind of production system. Paradoxically then, in many ways, simpler Lean systems require more maintenance than conventional systems. That is why they require a specific management system to sustain them.

How Can You Recognize Culture?

Remember that we can define culture in a work organization as the sum of its individuals' work habits. A related way to think of culture is that it is the knowledge an adult needs of how things are done to stay out of trouble as a member of a group. One of the interesting things about culture is that for group members, culture is invisible. It is the things that are given, or "the way we do things around here." It is typical not to question this kind of thing, or even to realize there are alternatives to it. Yet, it is easily possible to "see" work culture in a production environment by asking basic questions about common practices, such as:

1. Do those in similar positions follow the same sequence of steps?
2. What are inventory practices around here?
3. How often does management look at the status of production?
4. Who is involved in process improvement activities in this area?
5. What is the typical response when problems arise?

Asking these questions would reveal some of the distinctions between the cultures in conventional and Lean production environments.

The examples in Table 1.1 give a partial picture of the pervasiveness and everyday manifestations of culture in conventional and Lean production,

Table 1.1 Visible Attributes of Different Cultures in Mass and Lean Production

Cultural Attribute	Mass Production Culture	Lean Production Culture
Inventory practices	Managed by IT system Ordered by forecast Stored in warehouse areas or automated storage and retrieval facilities Held in bulk containers Moved by lift truck Many hours' worth or more per delivery Delivered by the skid or tub by forklift to vicinity of use	Managed visually Ordered to replenish actual use Stored in FIFO racks or grids addressed by part number Held in point-of-use containers Container quantity and number of containers specified per address Precise quantities (often less than an hour's worth) delivered to point of use Deliveries by hand cart or tugger
Production status	Checked at end of shift, beginning of next shift, or end of week Checked by supervisor, higher level managers	Checked by team leaders several times an hour Checked by supervisors four or more times a shift Checked by value stream managers once or twice during the shift Updated for all involved in tiered daily reviews of the previous day
Process improvement	Made by technical project teams Changes must be specifically "chartered" No changes between "official" projects	Can be and routinely are initiated by anybody, including operators Regular, structured vehicles encourage everyone from the floor on up to suggest improvements and perhaps get involved in implementation Improvement goes on more or less all the time, continuously
Problem solving	Do whatever it takes to hit today's numbers! Work around the problem; just meet the daily/weekly production/productivity goal	Record immediate circumstances of the miss, interruption, or breakdown Put temporary countermeasures in place if needed to serve the customer Assign task to identify and then to eliminate causes of problem

as well as how they differ from each other. In this view, culture is made up of myriad habits and practices that make it possible for all of us to go through the workday without having constantly to think about who, what, where, when, how, and why.

Culture allows us to operate more or less on autopilot during the workday. By the same token, a distinct culture also makes it easy to identify countercultural behaviors, practices, or events. In most groups, countercultural behaviors tend not to take root without many other climatic changes.

Overcoming Cultural Inertia

One implication of culture as a collection of habits and practices is that it has incredible inertia and momentum going for it. Cultural inertia is like a body in motion tending to stay in motion in the same direction unless acted on by an external force. Conventional production systems include a culture. So do Lean production systems. But, when you change production arrangements from batch or conventional to Lean, the culture does not make a similar change unless you take specific action to replace one management system with another. That's the parallel Lean implementation noted earlier, implementing the Lean management system.

Conventional habits and practices live on, even if the layout, roles, sequence, material, and information flows have changed. In one example, operators whose area switched from material requirements planning (MRP) schedules to pull signals were quite inventive in figuring out how to get access to a copy of the MRP schedule, which they then followed regardless of the pull signals. In this case, the fabrication operators regularly overproduced according to the discarded schedule they retrieved every day from a trash can near the dispatch office. It was not until they were found out, and the schedule paperwork began regularly to be shredded, that they had to follow the pull signals. Another common occurrence is for operators in newly converted flow lines transformed from batch build to go right on building. When the line fills up, it is typical to see the overproduction stacked on the floor, overstacked on conveyors, overflowing containers, or sitting as piles of unused documents, or as specifications "completed" before the full requirements were defined—and which now need to be reworked. Similarly, I have learned it is commonplace in hospitals for nurses to hide cases of supplies like dressings, IV components, gloves, even equipment like infusion pumps, above the suspended ceiling tiles so that when the inventory system fails, they are not caught short of critical materials.

New Settings with Old Habits Won't Work

It is typical to see supervisors and team leaders in a newly rearranged area rushing off here and there to chase parts or jump on to the line to run production. In some cases, it is nearly impossible to convince supervisors or team leaders to make the hourly entries on production tracking charts because they are "too busy" to get to this task. Then, once the tracking charts are actually filled out, it is not unusual to see them simply pile up on (or under) a supervisor's desk with no attention at all to the interruptions documented on the charts. If the schedule has been met, there is no interest in what is on that "paperwork." And if the schedule hasn't been met, there is "real work" to be done—no time to waste with these records of interruption! That will not get the schedule out or hit the target today, and in the old—and ingrained—culture, that is all that counts.

In conventional production, it is seen as important to be busy doing something directly, physically linked to production. Waiting for a production instruction card to arrive before starting to produce simply seems wrong. So does limiting the queue instead of putting customers on hold. Standing and waiting for the next piece to come down a progressive build line is definitely countercultural in the conventional batch production world. In such an environment, these interruptions in the rhythm of production are not considered to be valuable diagnostic information, signaling an abnormal condition in the production system, that is for sure! Relying on the reduced inventory of parts called for in a pull system seems sure to lead to stock-outs down the line. There is no perceived value in recording data that document the operation of the process. Action is what counts, and if it is based on gut feel and experience, it must be right, because "that's the way we get things done around here!"

These are only a few habits of thought, interpretation, and action that people absorb as part of the culture in a mass production environment. They are at clear variance with the kinds of habits and daily practices necessary for the precise and disciplined execution Lean systems need in order to meet their promise for productivity, quality, and ongoing improvement. Table 1.2 compares a few of the ways in which mass and Lean cultures differ. Many conventional production cultural practices are strikingly tied to long-standing ways of relating to others at work. In contrast, many Lean practices are related to disciplined adherence to defined processes.

Table 1.2 Differences in Habits and Practices between Batch and Lean Cultures

Mass Production: Personally Focused Work Practices	Lean Production: Process Focused Work Practices
Independent	Interdependent, closely linked
Self-paced work and breaks	Process-paced work, time as a discipline
"Leave me alone"	"I work as part of a team"
"I get my own parts and supplies"	In- and out-cycle work are separated and standardized
"We do whatever it takes to get the job done; I know whom I can rely on at crunch time"	There's a defined process or procedure for pretty much everything; follow the process
"I define my own methods"	Methods and procedures are standardized
Results are the focus, do whatever it takes	Process focus is the path to consistent results
"Improvement is someone else's job; it's not my responsibility"	Improvement is the job of everyone
"Maintenance takes care of the equipment when it breaks; it's not my responsibility"	Taking care of the equipment to minimize unplanned downtime is routine
Managed by the pay or bonus system	Managed by performance to expectations

How to Change Your Culture

We usually refer to changing habits with the word *break*, as in "that's a hard habit to break." Similarly, many talk about "kicking" habits. In each case, these words imply that changing habits is a one-time thing, a discontinuous step-change from one state to another, which once accomplished is an event that is irreversible, over and done with, no going back.

Many habits that come to mind are personal and physical in nature: smoking, nail biting, various forms of fidgeting—jingling pocket change, fiddling with an ID badge, a pen, or glasses, etc. At some level, each habit provides a form of comfort. We tend not to think of our work habits in these terms because many of them are part of the particular culture at work, and that is effectively invisible. Nevertheless, these habits arise because they bring a form of comfort, too. In a conversion to Lean production, some of these habits will be a hindrance, and some will be a help.

Consider these examples of management habits in conventional mass production operations, things you want to stop doing under Lean management:

- Keep a quantity of extra material stashed away at all times; you might need it.
- Jump onto the line or expedite parts when things slow down, or throw in more people; meet the schedule!
- Always reorder more than the actual need when handling shortages, just to be sure you get enough.
- Approach people who are waiting for the next task and ask them to get back to work.
- Use an informal gauge of queue size; always keep the line full in case something goes flooey.
- Always maintain a minimum 10 percent surplus labor and plenty of WIP; something could go wrong.

You can think of many more once you start to see work habits and practices as something you do without thinking about it. There is nothing wrong with habits as such. We need them to make the workday more efficient. What is important to remember is this: *work-related habits are just as difficult to change as personal habits!*

Extinguishing versus Breaking Habits

It is helpful to think in terms of the technical language from behavioral science used in connection with changing habits. The term is not *break*. Instead, psychologists use the term *extinguish* when talking about changing habits. Extinguish implies a process, something that occurs gradually over time, rather than an event producing a suddenly changed state. Because of that, extinguish also implies a change that can be reversed under certain conditions. Think of Smokey the Bear's rules: douse a campfire with water, stir the coals and turn them over, then douse again. If you do not follow Smokey's process (douse, stir, douse again), you run the risk that the fire can rekindle itself from the live embers you failed to extinguish.

So it is with habits. They linger, waiting for the right conditions to reemerge. We have seen this kind of thing just days following implementation of new Lean layouts. Here are some actual examples of old habits reasserting themselves in areas newly converted to Lean layouts—again, examples you want to avoid:

■ Build up some inventory.
■ Allow longer or extra breaks.
■ Send people off a balanced line to chase parts or do rework.
■ Start a task early, even if all the information is not available from the upstream department.
■ Work around the problem today and let tomorrow take care of itself.
■ Leave improvement to the "experts" rather than wasting time on employee suggestions.
■ Do not bother with the tracking charts—we never actually do anything about recurring problems anyway.

Make Sure You Don't Slip Back into These Old Habits

To sum it up, you do not need a different management system for Lean because it is so *complex* compared to what you have done before. You need it because Lean is so *different* from what you have done before. Many of the habits in your organization, as well as your own personal work habits, are likely to be incompatible with an effectively functioning Lean production environment. You have a conventional mass production management system and culture. You need a Lean management system and culture. Chapter 2 shows how to go about making that change.

Summary: Technical and Management Sides Need Each Other

Because Lean production is a system, it does not matter where implementation starts. Eventually you will get to all of the elements. But sequence does matter when implementing Lean technical elements and the management system. We have learned that technical change must almost always come before cultural change. Technical changes create the need for changed management practices. More than that, Lean management does not stand on its own. Without the improvements in process stability and flow from the application of Lean tools, production will continue to operate in an environment of daily crisis. In most cases, it is not meaningful or credible to propose production tracking or other Lean management elements in an environment of repeated interruptions, delays, defects, breakdowns, and shortages. You cannot track flow interrupters when there is no expectation for flow and no established basis for comparing actual versus expected production for a given interval of time. You cannot assess material replenishment performance without standard reorder points, lot sizes, and resupply times. (For low volume, custom or one-off production, or variable duration cases or procedures, see Case Study 4.5 in Chapter 4.)

So, start with the physical and procedural changes to the production process, but do not implement them by themselves. Just as changes to the management system do not stand by themselves, neither do technical and procedural changes. Each of these changes requires support from changes in the management system—the support of new management practices— to maintain integrity over time. If that is not a law of nature, it is darn close to it! Each time you implement an element of the Lean production system, implement the elements of the management system right along with it. The elements of the management system give you the tools that help you sustain the newly converted Lean process, close the loop on the process focus, and drive further improvement. Those elements and the conditions for successfully establishing Lean management are what this book is about.

Study Questions

1. Does your management system reinforce finding and eliminating problems, or does it reinforce working around problems to meet the schedule?
2. Are your processes people dependent or are they process dependent, that is, based on documented processes everyone understands, respects, and follows until the process is improved and changed?
3. Have you implemented the tools of Lean production only to find they didn't seem to work or didn't hold up over time? Why do you think that happened?
4. Are senior managers familiar with and prepared for the part they need to play in leading and sustaining Lean beyond initial statements of support?
5. Do line leaders closest to the floor routinely spend time (up to 50 percent of their time for manufacturing supervisors, less in administrative processes) on the floor observing the Lean process, talking with the people, reinforcing standards, and addressing problems as they come up?
6. Have you seen old habits reemerge and stymie a recently implemented Lean change?

Chapter 2

Lean Management System's Principal Elements

The Lean management system, like the Lean production system, consists of only a few principal elements. Lean management is a system in which the elements are interdependent (as in the Lean production system). That means all of the elements have to be present for the system to work. A third similarity with Lean production is that none of the elements are complex or complicated. And finally, Lean management and Lean production themselves are interdependent; one does not stand for long without the other.

So, what are the core elements of the Lean management system?

Chapters 3 to 5 each detail the attributes of the core elements. The point emphasized in this chapter is the way that the elements combine to form a system. An automotive analogy provides a helpful illustration (see Figure 2.1). Think of the three principal elements as representing major parts of a car: the engine, transmission, and controls. The fourth element is the fuel.

The Principal Elements of Lean Management

In Lean management, leader standard work (element 1) has the greatest leverage on maintaining the health of the Lean production system. It is the engine. (But, see Case Study 3.1 in Chapter 3.) Leader standard work is the first line of defense for Lean management's focus on the standardized Lean production process. When the leader follows his or her

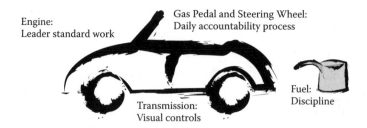

Figure 2.1   Making Lean management go.

standard work effectively, both Lean production and Lean management have a good chance of operating effectively. Remember, Lean management and Lean production are interdependent! One will rarely stand alone without the other. But when the Lean management system is healthy and operating effectively, Lean production will be healthy and operating effectively as well.

The transmission in Lean management is its visual controls (element 2). The visuals translate performance of every process into expected versus actual throughout the production and management systems. These data are recorded regularly and frequently, often many times a day, by those closest to the people doing the work. They are displayed in highly visual, widely accessible, readily reviewed formats, interpretable from 10 feet away. In these ways, the visuals convert the driving force of leader standard work into traction. Visuals give leaders the ability to quickly spot and move to action where actual performance has not met what was expected. And, visuals give frontline workers a way to signal clearly when things have gone wrong, when their work has been interrupted, delayed, or they have detected defects.

The controls in Lean management, the steering wheel and gas pedal, come with the daily accountability process (element 3). Through daily accountability the leader can steer, setting direction for improvement activity in the area; e.g., which of the gaps between expected and actual captured by the visual controls should we work on? Daily accountability also allows the leader to control the pace: how much gas to give for improvement, how quickly we should expect improvement to be completed, how many resources we should assign to this task.

At first blush, none of this sounds like a big deal to accomplish. Assembling a daily checklist for leaders certainly is not difficult. Whipping up a bunch of visual tracking charts is also a simple matter. Lots of people can produce those using Excel in no time at all. The basics for daily accountability merely involve scheduling the appropriate daily recurring 15-minute meetings: one for the team leaders to get together with their supervisor and the other for the supervisors and support group representatives to get together with the value stream manager. In fact, you could consider most of these tasks as fairly straightforward administrative details. Why should this merit discussion, much less a whole book? Because a critical fourth element is necessary to make the whole thing "go."

Discipline, leaders' discipline especially, is the fuel that powers the engine that makes the entire system go. Establishing leader standard work, visual controls, and a daily accountability session will amount to nothing without the discipline needed to execute these elements as designed and intended.

This is particularly true as you are starting out on your Lean journey. Remember, establishing new habits requires extinguishing the current competing ones. Learning new habits requires consistent positive reinforcement of the new and negative reinforcement of current habits already deeply rooted in place. This is a difficult course, one that experience shows is easier to avoid than to follow.

Lean Elements Need to Work Together

You will find there is little "coasting" in Lean management. If you stop following through on any of the three principal elements because things seem stable and in control, it is certain that you will soon face unstable and out-of-control processes that require you to reinstitute the very elements you thought you could do without. Remember the law of entropy, in which organized systems tend to move toward states of increasing disorganization? Lean production and Lean management are not physics, but the law of entropy nevertheless seems to apply. Ease off on focus or discipline and your Lean systems will quickly deteriorate along with their hoped for results.

Returning for a moment to the automotive analogy for the Lean management system, imagine you were fortunate enough to win a new car in a fundraising raffle. Consider your shock when you go to pick it up only to find an engine in a crate, a transmission sitting on a bench, a speedometer and steering wheel in boxes, and a can of gasoline. If the parts are not assembled and working together, they will not take you where you want to go (Case Study 2.1).

CASE STUDY 2.1: HOW LEAN MANAGEMENT FAILS WHEN THE ELEMENTS DON'T WORK TOGETHER

For the Lean management system to operate effectively, its elements must work together. In one case, leader standard work had been in place for team leaders and supervisors for nearly a year. The area in question produced several lines of upholstered chairs, most of which had been in production for many years. The chairs were made to order.

Demand for them had declined to the degree that they were only produced sporadically. On the occasion in question, a day's worth of production was called for on one of these aging lines. An experienced assembly operator from elsewhere in the seating department was assigned to build the day's order. He was not experienced with the specific line, but the supervisor had reason to be confident. She had an experienced chair builder, standardized work for the build process, and a team leader in the area.

As it turned out, the operator did little more than nod to the standardized work and then began building the chairs approximately according to the standard sequence. By failing to follow standardized work, he missed a critical step in the standard sequence of work elements that involved spraying glue onto the foam substrate to which upholstery fabric was then applied. This caused a known flaw once the product was in use in the field, a flaw the standardized work had been revised specifically to prevent.

The team leader's standard work called for verifying the operator's adherence to the production standardized work. The supervisor's standard work called for spot-checking the team leader's adherence to his standard work. Both were carrying their standard work documents with them that day, but neither exercised the degree of discipline to look beyond the surface. (Instead, the situation was probably assessed as: "The operator is in the workstation building the chairs and they look OK from here. The production tracking chart is being filled out and the pace is good.") What was required was to actually observe the operator following the standardized process, element by element, in sequence. Those sequenced elements called for using a template to mask off glue overspray from the troublesome areas.

Once the customer complaint was registered, it was clear beyond a doubt that the cause was failure to follow standardized work—on the part of the operator, team leader, and supervisor. Two layers of defense for the integrity of the process had failed because of coasting through the discipline of standard work. The checklists were in place and the boxes had checks in them, but the disciplined in-depth observation implied in leader standard work was missing. Engines, as a rule, do not function without fuel; leader standard work without discipline is no exception to this rule.

Execution Is Key to Lean Management

Creating checklists is easy. Developing and posting operator standard work is straightforward. Filling out production tracking charts is not a demanding routine. In the absence of leaders' disciplined adherence to their standard process, all of this is of no value in preventing a known defect from being produced and shipped. Not only do the elements of the Lean management system have to be in place, but also each has to be scrupulously observed for the system as a whole to work. Case Study 2.1 also illustrates, in a small way, the dependence between results from Lean production and an effectively functioning Lean management system. Lean management acts as the eyes and ears of discipline and diligence. Lean management monitors whether the Lean production system is being faithfully executed, and sounds the alarm when execution deviates from design.

On a more positive note, as you get better and better at executing the elements of Lean management, your job as an on-the-floor leader will become easier and easier, increasingly free of the "fire alarms" that you often chased all day long most days. In fact, one successful value stream leader spontaneously remarked to me that his area had become so stable and predictable that he now looked forward to uncovering flow interrupters and other abnormalities.

Those things gave him something "fun to focus on," he said, with the opportunity to lead a kaizen to attack the source of the problem, make his work more interesting, and make his area run that much better. His daily routine includes scrutinizing the daily production tracking data for ways to make improvements in the process. He has reduced the interval of observation for production tracking by a third (to 20 minutes) to be able to catch smaller interruptions. In a longer interval, these interruptions might not cause a miss, and thus would go unnoticed. His team leaders are Pareto-charting reasons for misses in two different areas, looking for the most frequent problems in order to attack them, and so on. This is an example of what Lean management can look like and the ways it can sustain and extend the gains in Lean production.

Implementing Lean Management: Where to Begin?

Leader standard work (detailed in Chapter 3) is the highest leverage element in the Lean management system, the engine that powers the system. But in most cases, leader standard work is not the element to start with when implementing Lean management. Here's why.

My understanding is this: the intent of leader standard work is to guarantee the integrity of the standardized Lean production process, whatever its nature, whatever the setting. A standardized Lean production process comes from applying the tools of Lean production to reduce the wastes of delay and interruption, improve stability and flow, and reduce frustrations. The Lean management system, as noted in Chapter 1, sustains Lean production and extends the gains from a Lean transformation. This formula implies a sequence in which the Lean production, with its standardized production process, comes first. So in most cases, lead with Lean production immediately followed by Lean management.

Even if you start Lean production with an improvement in one limited area, you will have improved that area's stability. People who lead production process of any sort in any kind of setting tend to be pragmatic. Show them something that works better, that makes the job a bit easier, reduces even a few recurring problems, and their reaction is likely to be positive. Now you can have a conversation about Lean management, which might go something like this:

You (Y): Pretty nice improvements in area X, eh?
Leader (L): Yes, that cluster of problems really seems to be gone, at least for now.
Y: But you know what's happened before, with other improvement projects, right?
L: Yeah, they don't seem to last no matter how much better the work goes.
Y: Well, this time we're going to try something new. It's simple. We use a production tracking chart [detailed in Chapter 4]. The chart shows

the number of widgets you expect the process to produce every hour [for example]. So, every hour the team leader writes in how many widgets were actually produced. When actual production falls short of what you expected, the team leader writes in what happened, in the form of what we wanted but didn't have, or what we had but didn't want. It really sharpens focus on the process as it cycles.

L: So the idea is to highlight problems? That *would* be something different!

Y: That's right. Remember, Lean is first about finding the problems, then about eliminating them.

L: So, once we capture the problems, we do something to keep them from creeping back?

Y: Exactly! Plus, you'll routinely be alerted to other interruptions and delays that might have been hidden by the ones you just addressed. Now that you've implemented visual controls—the production tracking chart—the next step is to take the chart to a brief daily stand-up task assignment meeting. This is an accountability meeting where you assign someone to look into each problem to find its cause, and then to suggest actions to eliminate it [detailed in Chapter 5].

L: So, is this what you mean by continuous improvement?

Y: Yes, this is a big part of it, but there's one more piece. The last piece belongs to you, your version of standardized work. It's like a checklist. The first item I'd suggest for your *leader* standard work is to stop by the area every hour or two to review the tracking chart and check on adherence to production standardized work. Are people following their standard work? Is the tracking chart being completed on time? Are problems clearly described? Has the team leader's response been appropriate? Is there something you need to do today, beyond the action the team leader took?

L: Like I'm making regular rounds to check up on the process, eh?

Y: You've got it—monitoring adherence to production standardized work and reviewing the production tracking chart. Those are the first items to go on your leader standard work. The next item for your standard work is to lead the daily accountability meeting [detailed in Chapter 5]. Completed accountability tasks often produce process improvements by changing how something is done. When that happens, you update your leader standard work to monitor the change. That way, your regular routine includes monitoring the new improvements to keep from backsliding into the way we did things before.

L (with some skepticism): OK, but why should I believe it'll stick this time?

Y: Fair question, but think about it this way. You and your team are now regularly keeping a detailed eye on the process. You're catching problems as they occur. You assign specific actions to a specific individual to complete by a specific date to get to the bottom of the problem and suggest how to eliminate it. Then, when you're already in the area, you monitor new changes to the process, making sure people understand them and that the changes are working as intended. When you do those things, you guarantee the integrity of the process and the improvements being made to it. Make sense to you?

L: You know, I think it's worth a try. Let's talk about that production tracking chart.

I have had many conversations like the one above. They boil down to applying Lean tools to make changes that improve the stability of a process, then showing leaders the straightforward structure, tools, and practices of Lean management that sustain the initial gains and surface other opportunities for improvement.

Most leaders understand their own interests. They readily see that a few easily mastered tools and practices can help their areas perform better and more effectively. And, at the same time, Lean management gradually frees leaders' time for further improving their areas. Standard Lean advice calls first for stabilizing a process, followed by standardizing it, then simplifying or improving it. Lean management not only sustains the initial gains from Lean production tools but also drives further improvement.

First, Improve Stability

But, until you demonstrate an improvement in stability of a process by applying the tools of Lean production, leader standard work comes across as a waste of time, a bureaucratic abstraction without real meaning. When this is the case, instead of helping, leader standard work is just one more administrative chore disconnected from the reality of daily chaos. Conventional line leaders have learned to be creative, to improvise in order to meet their daily goals. Improvising, as in working around daily problems, and standard work do not go together well.

Most of the organizations I have encountered that attempted to start their Lean implementation with leader standard work found it difficult, confusing, and ultimately unsuccessful. This difficulty makes sense when you think about the unstable state of many conventional operations before a Lean conversion begins. Leaders spend their days firefighting and expediting. In this kind of turbulent, interruption-prone environment, standardizing what a leader is supposed to do during the day is not going to be meaningful. Importantly, without an improvement in process stability, you cannot address the ever-present but rarely asked question—"What's in it for me?"—with a conversation like the one above. That is, you cannot make the connection between Lean management and sustaining—and extending—the benefits from the improvement in process stability if you have not first applied Lean production tools that produce improved process stability.

Start with Visual Controls

In most cases, technical Lean changes create or improve stability. Think of a production or order processing work cell with steady, paced work replacing the previous stop-wait-search-start condition. Or, consider a pull system that visually controls production or orders between operations, in contrast to a material requirements planning (MRP) push system in a plant or the lack of any systematic scheduling at all in an office process. In a Lean production system, you use capacity to produce what is needed. No more overproducing one component while the needed item is out of stock; no more drowning in unneeded supplier deliveries but still being short of what you need *right now*. With these stabilizing technical changes in place, the need for cultural change—for changed leadership practices and behaviors—comes into focus. It now makes sense to implement a Lean management system in order to sustain the new stability and identify opportunities to improve on it.

As suggested in the section above, begin your Lean management implementation by linking it to the technical Lean implementation, whether it is a kaizen or a complete value stream or support process makeover. As the last step in the technical implementation, put visual controls in place as described in Chapter 4. That is, pair your implementation of Lean management with your Lean production implementation. Now you have the elements of standard work for leaders that make sense in the leaders' own terms, sustaining the benefits from the Lean conversion. Those standard work elements might look like this:

1. Go to the visuals regularly to verify they are being completed consistently, in a timely manner, and with appropriate specificity. Understand problems where noted, verify appropriate action has been taken or initiate it.

2. Conduct the standard accountability meeting with the visual tracking charts or their data brought into the session.
   - Ask about the misses recorded on the visuals.
   - Make assignments to understand and act on the causes of process misses and system breakdowns, as well as assignments based on your and others' observations in the area.
   - Follow up on current and overdue assignments.
3. Add to leader standard work as appropriate to monitor changes from completed accountability tasks.

For most organizations the best advice is to begin with stability improving Lean projects and use them to prepare the ground for Lean management. With the area operating in a more stable state, the leader has time to pay attention to visual controls and to act on the feedback they provide on the health of the process. This closes the loop on process focus, sustains improvements from the Lean project, and drives further improvement (Figure 2.2). With these steps, the Lean management implementation grows from the technical Lean changes. The technical changes create the conditions in which Lean management makes sense to those asked to follow its new direction.

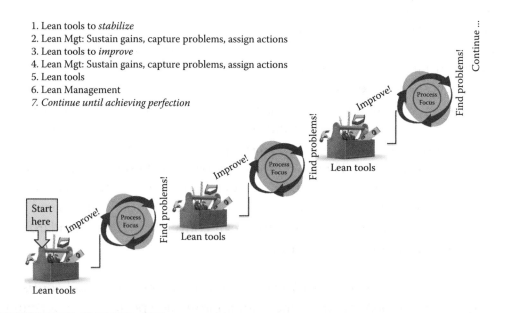

1. Lean tools to *stabilize*
2. Lean Mgt: Sustain gains, capture problems, assign actions
3. Lean tools to *improve*
4. Lean Mgt: Sustain gains, capture problems, assign actions
5. Lean tools
6. Lean Management
7. *Continue until achieving perfection*

Figure 2.2   Stabilizing and improving: Lean production and Lean management work together to produce continuous improvement.

When Implementing Leader Standard Work First Can Be Effective

I have seen the occasional situation where starting a Lean management implementation with leader standard work can be effective, in each case in process industry operations or highly automated discrete manufacturing environments. The distinguishing characteristic is the relative lack of repetitive hand labor in the manufacturing process.

Leader Standard Work in Automated Production Environments and Process Industries

In highly automated operations, relatively few operators tend multiple, often widely separated, pieces of high-volume automated production equipment. Those who are on the floor typically move from asset to asset, whether to stock them up, change components, check process parameters, make adjustments, perform routine in-process cleaning, maintenance, or inspection tasks, and so on. The few floor operators typically follow habitual (but usually not standardized and timed) routes. Some examples are operations that produce discrete units in large volumes, such as precision machined automotive components, web press printing or film extrusion operations, microelectronic devices, or alkaline batteries. Others are continuous process operations producing materials such as steel, chemical, or petrochemical products.

Floor Time for Floor Leaders

Just as with conventional discrete manufacturing operations, production leaders in conventional high-automation or process operations are often occupied off the floor with a variety of tasks, or distracted by them when on the floor. Much of what these floor leaders do was once the responsibility of support groups such as human resources, finance, product development, quality, production control, engineering, environmental compliance, or maintenance. In effect, other functional groups in the plant or business have over time shifted parts of what used to be their work to production floor leaders. Some examples are tasks related to payroll, attendance and FMLA leaves, inspections, calibration, cycle counting, product trials, discharge sampling, customer relations, and a variety of record keeping and reporting tasks.

Gradually, this work has come to occupy by far the majority of floor leaders' time. For example, it is not unusual for frontline leaders in continuous process operations to spend virtually all their time in control rooms or sitting in front of a computer monitor. The cumulative effect of shifting this work from support groups or other departments to frontline leaders typically goes unexamined. In case after case, the effect is to greatly diminish focus on and support for the value-adding work processes and the people in them.

In one high-volume, highly automated, discrete manufacturing plant that was an early adopter of Lean management, a young engineer serving as a frontline leader had been frustrated by the lack of process improvement activity in his area. He recognized that several hours of floor operators' time per 12-hour shift was occupied by, as everyone put it, "chair time." That is, operators would make their rounds performing the duties (such as stocking components, checking equipment calibration) required to keep production going, and return to sit in a chair in what was, in effect, an on-the-floor den they had created for themselves.

This young supervisor decided to stop doing some of the non-value-adding tasks for other departments so he could focus some of his time on process improvement. (He had the support of his plant manager based on the proposal to show this use of his time would result in improved process performance.) He spent time on the floor with his operators, first explaining himself (see below), to verify how long they took to do their tasks and the time it took to walk from one task to the next.

Finding Free Capacity

Here is what he found. The total for all four operators' task time plus walk time (the total operator work content plus travel time per shift) was 31 hours. The total planned available operator time per shift (12 hours minus breaks, lunch, etc., or 10.5 hours per operator) for the four operators was 42 hours. He and the operators realized they had, in effect, one extra operator.

When the supervisor first explained what he was doing, he made it clear nobody would lose their employment as a result of process improvements. Following through on this commitment, backed by his plant manager and following the plant's work rules, he converted one of the four operators— the one who was best at repairing and adjusting the complex equipment to keep it running—to be team leader. In exchange for giving up chair time, the other three operators now had someone to call on for help when production was interrupted by equipment problems, which had been a frequent source of frustration for them.

When the process was operating smoothly, the team leader had time to work on process improvements that previously had gone unrealized. The improvements were too minor to make the to-do list for maintenance or engineering, and nobody else had had time for them. Plus, the supervisor was more available to the people on the floor, time freed from what he used to spend working on tasks sent from other plant and company departments.

When I followed up, I learned the changes had gone relatively smoothly. Operators liked having a resource available to them right when they needed it to get a shut-down piece of in-line, complex electromechanical equipment back in operation. Operators had quickly gotten used to the new routes and duties they were asked to perform. There was a new position, a step up from production available for those who could qualify. And the supervisor reported substantial process improvements from the recently created team leader position.

Impact on Workers

It is important to state upfront the commitment that nobody will lose employment as a result of process improvements when that is true. Some people may be asked to do other jobs, but will not find themselves out of work. Might operators in this example have grumbled about losing some of their chair time? Sure, but they were well treated overall by the plant and parent company, and they received something in return: help on the floor when they needed it, and better access on the floor to their supervisor as well.

Increasing competition is a fact of life for organizations of all kinds. People recognize this, even if reluctantly, and understand that process improvements strengthen the competitive position of their workplace—especially when their employer treats them with respect. It makes sense for organizations to invest in process improvements, especially those smaller opportunities that are hidden in plain view and inexpensive to realize.

Standard Lean advice, which I endorse, is to invest 5 percent of the labor savings from process improvement in further process improvement resources, whether you call it a Lean team, continuous improvement positions, or coaching in Lean and other skills development for those in frontline leadership positions. It helps the organization's performance and provides more meaningful work for those involved. As a bonus in the case of newly created process improvement positions, it provides a vehicle for identifying talented people from the floor for potentially greater responsibility, most of whom, in my experience, would otherwise not have been noticed, much less considered for promotion.

Does Lean Management Apply in Process Industries?

The Lean management system identifies a set of practices and behaviors to close the loop on process focus and drive improvement. Process focus in a discrete production manufacturing or office environment assumes comparing actual versus expected production of discrete units of output. The key measures of process focus are based on takt time as the pace for production. These two assumptions—discrete production and takt pace—work well in repetitive production environments producing discrete units of output, whether in factory or office settings. These assumptions, however, do not hold in process industries. If the assumptions do not fit, do Lean and Lean management apply in process industries?

For example, consider continuous (24 × 7 × 365) process operations such as a steel mill or a chemical operation producing synthetic petroleum. In either case, it is certainly possible to measure actual versus expected output in tons or barrels produced. And of course operations like these use these measures, as much as indicators of process as of results. "How many tons or barrels did we produce today?" is a natural question to ask. What if the answer comes back as less than planned, required, or expected for any period of observation? In a process operation, as in discrete production, less than expected output raises the question: "Why? What happened that cost us production?"

Asset Focus in Continuous Processes

In a discrete production operation, you can usually answer this kind of question in terms of flow interrupters in assembly or problems upstream that starved later production steps. Investigation might take you to breakdowns in fabrication, problems in a finishing operation or with incoming materials, components out of spec, or operator performance issues (Case Study 2.2).

In a process operation, assuming reasonable control of incoming material, the problems are much more likely to be related to the process assets or breakdowns in process safety management. In a steel mill, the shroud may have misaligned, exposing the melt to the contaminating atmosphere as it was being poured into the continuous caster. In a chemical plant, the problem may have been equipment availability, or unplanned downtime, as either a measure of process or result. Both failure to operate the assets appropriately (causing unplanned downtime) and failure to adhere to process safety management procedures could be reflected in measures of process performance. Or unplanned equipment downtime could be because planned maintenance activities were not completed in the allotted standard time. And that could be because operations did not turn the assets over to maintenance on time. Or it might be because the right parts were not available at the right place and time for the maintenance techs to do their jobs.

CASE STUDY 2.2: PROCESS FOCUS BEYOND TAKT TIME

An automotive fuel system plant was committed to going Lean. Its production process had to meet tight tolerances at high volumes. It was a capital-intensive, precision machining operation that ran around the clock, five days a week. Demand varied by the needs of the automakers' engine plants, its customers. The automakers released monthly schedules to the fuel system plant, with level weekly requirements. So, demand was known a month in advance and was stable within any given week.

The fuel systems plant's first foray into Lean was a Total Productive Maintenance (TPM) program focused on reducing machine downtime. Breakdowns had been a chronic problem in this operation, and they became the focus of the new Lean initiative. TPM took hold and the glaring equipment reliability problems were progressively resolved. As equipment reliability stabilized, the plant's Lean team and operations leaders recognized there were far fewer major equipment breakdowns, but the plant was still losing production time.

Their next step was to further tighten their process focus. They began measuring overall equipment effectiveness (OEE) for their key production equipment. OEE measurement is based on tracking the so-called six big losses in OEE: minor stoppages, major stoppages, waiting for operator, waiting for material, yield or defects, and speed losses. The data are used to compute the OEE coefficient (see glossary).

The plant began tracking losses in each of the six categories every hour for each of its critical assets. Not only did OEE give a clearer picture of how production was being lost, but it also anticipated the questions a purely takt-based measure would prompt. That is, a takt-based measure would lead you to ask why we lost production last hour. In this automated equipment-intensive environment, the answer would almost certainly involve one of the six big losses.

Manager or supervisor: *Why did we lose production this hour?*
Supervisor or team leader: *The finish machining cell was down.*
Manager or supervisor: *Why was the cell down?*

The answer, which would prompt further investigation, would likely be a major or minor stoppage, or waiting for material or an operator. With OEE six losses monitoring in place, the questions can begin with: "Why were we waiting for material?" No backtracking is needed to uncover the details of a problem that happened yesterday or on the previous shift. The data are right there, recorded and preserved on the OEE tracking chart.

Process Focus and Leader Standard Work in Process Production

So how might leader standard work apply in continuous process operations, and when might it come before the other elements in Lean management, or even before a focus on standardizing operators' routine procedures?

Set aside maintenance-related concerns for this discussion. Maintenance issues in process industries—apart from the differences in environments in which maintenance is performed—are not that different from those in other industrial settings. These issues, such as access to assets, having the needed information, tools, parts, materials, and equipment, standardized work for maintenance procedures, and so on, are largely the same in process and discrete manufacturing.

In continuous process manufacturing, as in the high-volume discrete manufacturing example earlier in this chapter, there are usually few people directly involved in process operations. In addition to the high level of automation, continuous process operations are often characterized by material and process sensitivity, high temperatures and pressures, toxicity of the process, or all of these factors. As in the earlier example, operators often have defined, or at least habitual, routes through the plant with specific tasks or procedures to perform at each stop. In many continuous process industries, adherence to the operating and process safety procedures, no matter how vaguely defined or documented, often can be of critical importance in preventing catastrophic failures, not to mention destroying equipment and putting lives at risk.

The intent of leader standard work is to guarantee the integrity of the standardized Lean production process. Paradoxically, I have seen leader standard work applied as a first step in a process industry Lean implementation where operators' procedures could hardly be called standardized. Instead, the procedures were vaguely or only generally defined, received little, if any, attention, and were often out of date or obsolete altogether.

In this case, the organization was committed to a vigorous Lean production implementation but was only getting started. (See Case Study 8.4 in Chapter 8 for a description of this situation.) Recent catastrophic breakdowns in the plant made it clear that critical elements of the operation were out of control. It was in this context that leader standard work was developed to serve as an emergency stopgap measure and a diagnostic for which areas first needed Lean attention. Without the discipline of a committed Lean follow-up, leader standard work as a stopgap would almost certainly have faded into check-the-box ineffectiveness, but fortunately that was not the case.

When Leader Standard Work Can Come First

Where safety and operating procedures are not well defined, or have been left for operators to interpret on their own with little or no input from first-line managers, establishing leader standard work can make sense in advance of other aspects of Lean or Lean management. Here, the standard work elements called out critical parameters at each stop on an operator's route. Where the leader's observations met expectations, she could move on. Where the observation indicated a process, procedure, or setting out of compliance, the leader first took immediate action to correct the situation. She then followed up to clarify or update expectations, ensured each operator demonstrated how to do what was required and its importance, and monitored adherence on her subsequent rounds on the floor.

In high-volume, equipment-intensive automated or process industries, close focus on process means close focus on equipment and close focus on adherence to well-defined operating procedures. In these cases, takt-based measures are actually further from the process, further from the sources of variation than a measure like overall equipment effectiveness (OEE), timely preventive maintenance completion, or scores on procedure compliance assessments and process safety management. Lean management is about close focus on process to highlight misses and drive improvement. This principle operates as clearly in process industries as in discrete production environments, administrative, transactional, and healthcare settings.

A process does not have to resemble an automotive assembly operation or other industrial setting to benefit from a Lean transformation, and to be sustained by implementing Lean management. Every Lean application is an invention, one you create based on a small set of Lean principles: understand value from the customer's perspective, find the value stream that produces that value, focus on the processes in the value stream to identify and eliminate sources of waste from it, and strive to make the value stream responsive to your customer's demands. Then, the key is to keep repeating these steps until non-value-adding steps are completely eliminated.

You can come to an understanding of what process focus means in your operation from these few Lean principles and your in-depth understanding of your operation. The key ingredients are knowing what makes your process tick and your judgment. Then, it is a matter of converting process focus into improvement no matter what the nature of your process. Nobody can do it for you, but fortunately you are in the best position to focus on process and use it to drive improvement in your operation.

So, gas up the Lean Management Express and take it for a spin. You will find the road in front of you gets smoother and takes you to places you could previously only dream about.

Summary: Four Principal Elements of Lean Management

The Lean management system consists of four principal elements:

1. Leader standard work
2. Visual controls
3. Daily accountability process
4. Leadership discipline

Even considering its additional supporting elements, Lean management is not a complex system. In this respect, it is similar to the Lean production system; a handful of principles define both approaches. The principal elements are illustrated in depth in Chapters 3 to 5. Supporting elements are the subject of Chapters 6, 7, and 9 through 11. Lean management assessment instruments are included in Chapter 12 and Appendices A and B. Specific advice for engaging executives in your Lean initiative is in Chapter 8.

Another similarity between Lean production and Lean management is the high level of interdependence among the elements in each one. Not only must the elements be put in place, but they also need careful, daily attention. Left untended, the management system quickly deteriorates and loses its effectiveness, just as with the production system. But, when Lean management becomes "the way we do things around here," the benefits are considerable. Lean management that is well and consistently implemented helps bring the foundation of stability to Lean production conversions, a foundation on which ongoing improvements can be built. Establishing and building on this foundation are the subjects for the remainder of the book.

Study Questions

1. Have you seen Lean management tools implemented before technical Lean tools were deployed? Were the Lean management tools effective? Did they last? Why?

2. Are there visual tracking tools in your operation? How effective are they, and why?

3. To what degree are frontline leaders able to concentrate on the way the floor is operating; that is, about what percentage of their time are they able to devote to the floor? What is the effect on operations?

4. In the various improvement programs you've seen, to what degree do you find people have chosen one or two tools and ignored the rest? How has that worked?

5. Highly automated production processes are often said to more or less "run themselves." In your experience, does that seem to be true?

6. What, if anything, did you find surprising in this chapter?

Chapter 3

Standard Work for Leaders

Standard work for leaders, the engine of Lean management, is the highest leverage tool in the Lean management system. As mentioned in Chapter 2, leader standard work is the first principal element of Lean management; this chapter describes it in detail.

Leader standard work provides a structure and routine that helps leaders shift from a sole focus on results to a dual focus on process plus results. This change in focus is crucial to the success of a Lean operation. Moreover, it is perhaps the most difficult thing to accomplish in a leader's personal conversion from conventional or batch and queue to Lean thinking. Leader standard work aids this conversion by translating the focus on process, an abstract concept, into concrete expectations for the leader's own specific job performance. Just as standard work elements in a production workstation provide a clear and unambiguous statement of expectations, the same is true of standard work for leaders. The main difference is that virtually all of an operator's time at work is defined by standardized work. For team leaders, the proportion is still about 80 percent. The proportion drops to about half for supervisors and about a quarter for value stream leaders. In administrative and technical-professional workplaces, the percentages for supervisors, managers, and directors may be lower based on the variability in the work and rate at which units of work are completed.

Keep in mind as you consider the potential leverage of leader standard work that it may not be the element with which to begin your Lean management implementation, as noted in Chapter 2.

Leader Standard Work Is Process Dependent

Leader standard work also provides a foundation for continuity in Lean management in a unit. Each time a new team leader or supervisor starts work in a Lean area, things should continue to operate much as they have, assuming the process has been in a satisfactory and stable state. In batch operations, in contrast, one often sees a "new sheriff in town" mentality accompanying change from one supervisor to the next. That is, in these circumstances the management system depends on the person. With leader standard work, the Lean management system is process dependent, not person dependent. Instead, key aspects of the management system are captured and presented in leader standard work as a well-defined process in which core tasks and routines are explicitly called out.

There are several benefits to this approach. One benefit is continuity of basic practices across changes in incumbents, which minimizes variability that might destabilize the production process. But perhaps more important is a second benefit, especially for organizations involved in a transformation from conventional to disciplined Lean operations.

A Recipe for Success

This second benefit is that leader standard work quickly allows an organization to raise the game of the existing leadership staff, and highlight those unable to make the transition. Leader standard work does this by presenting a clearly stated recipe for success—the standards for expected behaviors for leaders in a newly Lean environment. This is superior to the alternative of waiting and hoping for so-and-so to "come around" or to "get it." Leader standard work is first focused on "doing it," rather than "getting it." It allows a much speedier separation of those who are willing and able from those who are not. Experience in Lean conversions suggests between 10 and 20 percent of leaders are unable or unwilling to make the transition. Glossing over certain leaders' failure to understand or support the Lean initiative carries a high risk of seriously slowing or degrading the effectiveness of a Lean conversion. Leader standard work makes these failures clearer, sooner. A Lean initiative is like any other program in raising questions about management's commitment to back up its words with actions. Management can act to address these questions more quickly based on the clearly documented expectations in leader standard work.

As the transition proceeds, leader standard work captures the cumulative to date essence of an organization's best practices in Lean management. This provides a solid starting point for those in leadership positions (See Case Studies 3.1 and 3.2, contrasted with Case Study 3.3). As such, standard work for leaders provides a leg up for leaders to improve their performance, building on the preserved experience of others. In this way, leader standard work is a specific means through which average leaders can consistently turn in above-average performance. If you think of the Lean management system as a kit of parts, leader standard work provides the instructions for how they fit together. It reduces ambiguity and sets the conditions under which an individual leader's success is more likely.

CASE STUDY 3.1: WHEN LEADER STANDARD WORK IS MEANINGFUL

A precision machining plant had been transitioning to Lean production for two years. With Lean production working smoothly, the plant added a focus on Lean management. The plant had visual tracking charts and was ready to begin with leader standard work. But based on past experience with Lean implementation, it paused.

In its transition to Lean production, the plant had learned not to just charge ahead with Lean production tools. The lesson came when it eliminated batch production of work-in-process inventory, instead implementing a pull (kanban or just-in-time inventory) system. It quickly found that key pieces of production equipment were unreliable. The reliability problems had been covered by inventory, and with the inventory gone, breakdowns caused repeated production stoppages. The plant rebuilt inventory while implementing an aggressive preventive maintenance program. The second try at using pull systems was successful. Lesson learned: First map the value stream to find the problems, eliminate them, and then go ahead.

When the plant turned to leader standard work, it first took a step back to discover potential problems. The idea behind leader standard work was for supervisors to spend about half of their time on the floor as the first line of defense for the recently stabilized and standardized Lean processes. To find potential problems, the plant asked: How do supervisors currently spend their time? To grasp that reality, the plant asked supervisors to keep time diaries for several weeks. That is, supervisors recorded each of their activities, its duration, and when they occurred, all day every day for several weeks during a six-week sample period.

The results were shocking: supervisors spent hardly any time actually supervising on the floor, unless in response to a crisis situation. Instead, most of their time went to tasks that had been "outsourced" to the supervisors from other parts of the company, whether corporate departments or plant groups in HR, payroll, training, quality, scheduling, R&D, product process and facilities engineering, health and safety, environmental compliance, accounting, maintenance, shipping, customer service, local government relations, and so on.

Wisely, before asking the supervisors to use standardized work, the plant acted to greatly reduce the time supervisors spent on activities that took them off the floor or distracted them when on the floor. The activities no longer of use to anyone were dropped. Much of the outsourced work was transferred back to departments where it came from. Some went to a new clerical position added to free the supervisors for value-added work. When leader standard work was introduced, the supervisors found it to be both realistic and meaningful. They had the time to follow it, and it provided a helpful vehicle for doing the job for which they originally applied.

CASE STUDY 3.2: WHEN LEADER STANDARD WORK IS NOT MEANINGFUL

In a not so happy example, the Lean team in an aerospace defense plant developed standard work for leaders in a careful, analytical process based on what they thought supervisors should do in a Lean operation. I had been invited to the plant to lead a workshop shortly after the leader standard work had been introduced. In the workshop, I indicated a first-line manufacturing supervisor should nominally spend about 50 percent of his or her time on the floor.

A frustrated supervisor in the back of the roomful of people stood up and heatedly said that in his typical 12- to 14-hour days, with all the corporate, engineering, scheduling, supplier, customer, accounting, HR, and other demands on him, he was lucky if in a day he had 30 minutes on the floor to see for himself how the work on his new product was progressing and what he might need to do to support it.

I learned during the course of the day that virtually all those who had been given the Lean team's version of leader standard work found it to be completely unrealistic given their present duties and demands on their time. The team's version of leader standard work was not meaningful to the leaders who received it. It failed to reflect their day-to-day reality, failed to help them improve their own and their area's performance, and failed to make for a better day at work. Understandably, to a person, the supervisors ignored the newly imposed leader standard work. After all, nothing had changed in the daily demands they experienced, demands that were not represented in the staff group's version of standard work.

CASE STUDY 3.3:   THE VALUE PROPOSITION FOR LEADER STANDARD WORK

A natural resources mining and processing company asked for advice implementing Lean. It had started its Lean implementation with leader standard work, and it had not been successful. That was where the company was looking for help.

The director of process improvement and one of his staff had thoroughly debriefed the supervisors in a processing unit that was spread out over a large area. (The unit's leadership team had agreed to experiment with the use of supervisor standard work.) From the debriefing interviews, the two process improvement people captured supervisors' regular activities, daily, weekly, and monthly. The two carefully analyzed the data accumulated from the interviews to identify what was common among the supervisors. From that, they developed a draft of supervisor standard work. The unit's supervisors had followed the draft standard work for three months, but had stopped using it about six months before my visit to the site.

I was interested to learn about their experience, particularly given the sequence with which the company had gone about its Lean implementation. It seems to me that Lean production implementations generally should come *before* implementing Lean management. In part, that's because Lean management should represent a value proposition for leaders. The value proposition goes like this: Lean production tools improve the process; Lean management tools sustain and extend the gains from Lean production. Process improvements usually mean a better day at work for supervisors (and frontline workers) by improving process stability and reducing interruptions, delays, and frustrations. Without those benefits, where is the value for line leaders? What is the rationale for asking leaders to adopt Lean management's behaviors, practices, and tools, including leader standard work?

We went to the processing area where supervisors had experimented with leader standard work, and talked with one of the supervisors who had been involved. From a dusty lower shelf she pulled out the binder with the leader standard work documents to show us the filled-out pages. "We used it for a few months," she said, "but then we decided to stop." Asked why, the supervisor said it really wasn't helping them do their jobs, except for one thing. She said the standard work reminded her to go to a remote part of the unit that the supervisors often forgot about. But that was the only benefit she could name, and it was not nearly enough to justify carrying and filling out the standard work forms, which after all were just a list of the things the supervisors did as a matter of routine.

Without the benefits from sustaining an improved process, the tools, practices, and behaviors of Lean management are just not meaningful to frontline leaders trying to keep their operation running from one day to the next. But when a Lean production application improves an area's stability, that's worth doing something to sustain. That is when Lean management becomes meaningful, when it helps keep a good thing going.

Leader Standard Work as Interlocking Layers

Leader standard work is layered with a degree of redundancy built in, linking the layers. Think of a job responsibility for team leaders, such as filling out the production tracking chart every 20 minutes and noting reasons for misses when the team misses the goal for that pitch or interval of time. Here's how the layers work for this task:

- The team leaders' standard work specifies this task.
- The supervisor's standard work calls for spot-checking the pitch, cycle, or production tracking chart at regular intervals several times a day (more frequently for processes that cycle quickly), initialing it each time. Further, the supervisor's standard work calls for leading a brief daily meeting of his or her team leaders to review the previous day's pitch charts to understand any misses in performance and ensure action has been initiated as appropriate. These are all consistent reinforcements for team leaders to focus on their processes.
- The value stream manager's standard work calls for initialing every production tracking chart in the value stream once a day, and leading the top tier of the daily accountability process every day. A key feature of these meetings is a review of yesterday's production tracking charts. The value stream manager scrutinizes the reasons for misses on the charts. The supervisor knows to be prepared to explain what happened.
- When appropriate (typically where the supervisor needs more resources or more encouragement to resolve the problem), the value stream manager assigns follow-up action items to the supervisor or support group representatives and posts the assignments on a visual daily task accountability board for review the next day.
- All of this is prompted by the specific requirements in leaders' standard work from value stream manager to team leader.

Could any of these three levels of leaders find ambiguity in the requirement to maintain and monitor production tracking charts, paying careful attention to reasons for missed pitches, and initiating appropriate corrective action? It is unlikely, because these steps appear as routine daily items in their standard work (see Table 3.1). With this kind of readily audited daily direction, establishing focus on process is simply a step-by-step routine. It makes Lean accessible and actionable to even the most inexperienced leader and to the most unreconstructed conventional or batch thinker.

Leader Standard Work Shows What to Do—and What Not to Do

Case Study 3.4 shows how standard work can help a leader see what needs to be done. Standard work for leaders also functions in the converse; that is, it can also show what should *not* be done. Case Study 3.5 illustrates this point.

Table 3.1   Typical Items in Leader Standard Work

Frequency	Team Leaders (TL)	Supervisors (Supe)	Value Stream Mgrs (VSM)	Plant Manager (PM), Execs (When in Plant)
Once daily, typically repeated each day (or each occasion for plant managers and executives)	Check call-ins Adjust labor plan	Shift change coordination Daily admin tasks	Daily admin tasks	Review performance trend charts
	Lead team start-up (tier 1) meeting (5–10 min)	Attend a TL start up meeting	Night shift gemba walk	Spot check, sign-off pitch charts, other visual controls
	Floor check production start-up	Floor check production start up	Lead value stream task/ improvement (tier 3) meeting (10–20 min)	Lead weekly plant performance/ improvement review meeting (PM)
	Supe-TLs (tier 2) meeting (5–15 min)	Lead (tier 2) meeting w/ TLs • Misses, issues, improvements • Daily task board due and new items	Daily gemba walk w/ one supe	Spot-review process and product improvement work
	Gemba walk w/ supervisor	Attend weekly recurring plant-level meetings	Formal audit of one area	Verify leaders' standard work
	Supe-TLs meet accountability, improvement (5–15 min)	Gemba walk w/ TLs one on one	Attend weekly recurring plant-level meetings	Verify TL, supe on floor or why not?
	Daily (weekly) continuous improvement meeting w/team	Spot-check buzzer-to-buzzer work		

Table 3.1 (*Continued*) Typical Items in Leader Standard Work

Frequency	Team Leaders (TL)	Supervisors (Supe)	Value Stream Mgrs (VSM)	Plant Manager (PM), Execs (When in Plant)
	Next day planning • Labor plan • Prep for team start-up meeting	Spot-check, sign-off each pitch chart Review status of all other visuals		Gemba walk each VSM, staff manager weekly (PM)
Many times daily, often specified by time of day or number of times	Monitor buzzer to buzzer work before, after breaks	Spot check standardized work in each TL's area	Spot-check buzzer-to-buzzer work	Floor time
	Verify pitch by performance • Record reasons for variation • Note, act on flow interrupters	Floor time	Spot-check, sign off pitch charting in each department. Spot-check other visuals	
	Monitor standardized work in each station • Check compliance • Reinforce, correct performance as needed		Spot-check standardized work in each department	
	Revise production work standard as needed		Floor time	
	Train operators as needed			

CASE STUDY 3.4: LEADER STANDARD WORK'S ROLE IN CREATING AND MAINTAINING STABILITY

An illustration comes from the case of a supervisor's return from vacation. He had been using standard work to structure his workday routine for several months before leaving for a few days off. His area, a subassembly and final assembly operation with about 35 people, was running smoothly when he left. Upon his return, he discovered instability and chaos. He found purchased part shortages, shortages of some parts made in-house, and other in-house parts out of specification and unfit for use. The area was behind schedule, and the team leaders had been spending all of their time trying to expedite the various missing parts.

The supervisor immediately jumped into the fray, joining the team leaders in expediting, calling suppliers, and going to his in-house suppliers to attempt solutions to the out-of-spec problems. The assemblers continued to build whatever they could, robbing parts from one order to complete another, offloading partially built units wherever there was space on the floor, and generally keeping as busy as they could doing whatever work presented itself no matter when it was due out.

After two days of exhausting but largely fruitless efforts, his area remained behind schedule and in a state of jumbled confusion. At that point, his value stream manager, who had been observing the situation, suggested to the supervisor that he might try returning to his standardized work. The supervisor listened and beginning the next morning took his boss's advice. Instead of running here and there, he and his team leaders stayed with their process. As shortages interrupted production, they directed the crew members to other tasks while documenting what had caused the stoppage. They transmitted this information to the appropriate support groups, in this case engineering and materials management as well as the value stream leader. These situations were also topics at the value stream's daily accountability meeting. Within a day, the missing parts began to be delivered and the problem with out-of-tolerance components was resolved. The assemblers concentrated on completing the oldest orders first and were on their way to working up to the current day's schedule. A sense of calm was restored to the area.

By the time I saw the supervisor, four days after he returned from vacation, he swore he would never deviate again from his standardized work. At the same time, he acknowledged the powerful pull of past practices (supervisor as firefighter), though he was also struck by the comparison in effectiveness of the old and new approaches. He has been a faithful adherent of standard work, for both himself and his team leaders, ever since.

CASE STUDY 3.5: THE VALUE OF STICKING TO THE PLAN

A new team leader had been on the job only a few weeks and had no precedents to follow. Her area, which covered two assembly lines that were building two different sizes of the same family of products, had not had team leaders before. She was the first.

The area had been well and carefully designed by a bright, technically oriented value stream leader and his staff. Standard work was posted for all the area's production-related operations, in-cycle and out-cycle as well. Even so, there were repeated interruptions, usually because in-house or purchased parts were not available when needed, or because the computer algorithm used to schedule the finishing processes was unable to produce an even flow of units into assembly.

I had encouraged the value stream manager to establish standard work for his new team leader and later suggested some edits to a draft he had produced. A few weeks later, I saw her in her production area and asked how it was going with her standard work. She pulled the page of standard work out of her back pocket and held it up in front of her like a school crossing guard with a stop sign. "It's great," she exclaimed, going on to say it was useful in reminding her what she needed to do. But better than that, she added, it was a terrific way to explain to the many people making requests of her why she could not do what they asked. "I have to stick to my standardized work," she said she told them.

Since then, more team leaders have been added to her value stream, each with well-defined standard work. Now, every operator in the value stream has someone whose standard work includes responding to his or her questions and requests.

Leader Standard Work Should Be Layered from the Bottom Up

The most important activity in a production operation is production. In a Lean environment, carefully designed and monitored processes define production activity. When the process operates as designed (and refined), it meets its goals for safety, quality, delivery, and cost.

One of the two primary responsibilities of leaders in a Lean production environment is to see to it that the processes run as designed. (The second is to improve the processes.) The process is most likely to run as designed when operators doing the actual production and related jobs are following their standardized work. If they are, things should run predictably. Again, the intent of standard work for leaders is to guarantee the integrity of the standardized Lean production process.

For these reasons, standard work for leaders is often built, or layered, from the bottom up, usually based on standard work or standard procedures at the value-adding level (Case Study 3.3) (but for an exception, see Case Study 8.4 in Chapter 8). These are examples of leader standard work built from the bottom up:

- Develop team leaders' standard work based on maintaining production at takt time (or the equivalent takt-derived pace where available) and ensuring that standard work is being followed in the production process.
- Build supervisors' standard work based on monitoring and supporting team leaders in carrying out the responsibilities in their standardized work.
- Similarly, build value stream managers' standardized work to monitor and support supervisors for following the responsibilities in their standard work.

Executives should also have standard work for their time on the production floor, but with a different focus. Their production floor standard work should usually focus on the health of the Lean management system (see Chapter 8). Their focus on Lean management is based on this proposition: if the Lean management system is healthy, then the Lean production system will be healthy as well. Production floor leader standard work for operations and other executives interested in the nuts and bolts of Lean production (which is unusual) can also include elements to verify that the chain of standard work is being upheld and the production process is being supported for stability and improvement.

Designed Overlap in Leader Standard Work

This network of support for the integrity of the production process is represented in Figure 3.1. The overlapping links—leader standard work—are the mechanism by which the Lean management system supports the integrity of the standardized Lean production process, keeping it from sinking into the dreaded pit of instability, backsliding, and despair. Each of these links is discussed in more detail in the next section of this chapter, and for executives in Chapter 8.

The interlocking and somewhat overlapping set of leaders' standard work is directly comparable to the quality checks built into successive workstations on a Lean assembly line, process steps in order management or technical-professional processes, or standardized procedures in a hospital. On the line, the standard work elements for each station typically include checks on the quality of some aspect of the work performed in the previous station. These redundant quality checks are a strength in a Lean process. They often contribute to eliminating the need for a separate inspection position in which no value-added work is performed. And so it is with leader standard work. Each successive level of leader standard work includes checks for the tasks that culminate in supporting the integrity of the most important work in the operation—the standard execution of the production process. That is the purpose of leader standard work: maintaining the integrity of the standard Lean process, whether in an office, a factory, a healthcare facility, or any Lean process.

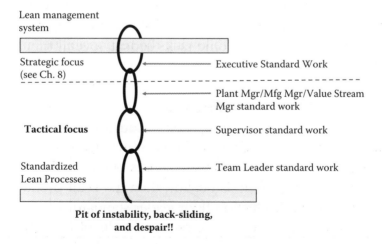

Figure 3.1   Overlapping elements in leader standard work across levels supports the integrity of the standardized Lean production process.

What Does Leader Standard Work Cover?

Most of a frontline worker's day—over 95 percent of his or her time—is accounted for by standardized work or standardized procedures. In a Lean environment where work is paced by takt time, most production and related work is timed and balanced. Leaders' work is rarely timed this closely. Even so, it is important to pay attention to the total work content called for in leaders' standard work to be sure they can perform it effectively and thoroughly. In positions at increasing distance from production, leaders' standard work usually becomes less structured. And as organizational level increases, less time is specified in standard work with more time for discretionary tasks. In the same way, as organizational level increases, fewer elements need to be performed in a specific sequence or at specific times of day.

Leaders' standard work includes coverage of visual controls and executing the daily accountability process. That is the source of its high leverage. Follow leader standard work and you maintain the principal elements of the Lean management system. Maintain the Lean management system and you sustain the gains from Lean production, but that's not all. A Lean management system allows you to capture further process misses (with visual controls), convert them to improvements (in the accountability process), and sustain the new gains (by updated leader standard work). Lean is fundamentally an improvement system; leader standard work sustains and extends improvement (Figure 3.2).

Leader standard work includes some tasks that are specifically sequenced to happen at indicated times. Others occur once a day, once a week, or as the need arises. Some tasks repeat several times every day.

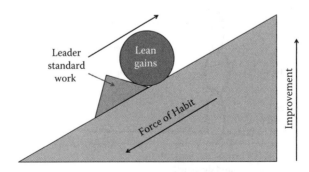

Figure 3.2 Leader standard work sustains and extends Lean's gains by closing the loop on process focus, the key to sustaining gains from Lean production. (Courtesy of Scott Simmons.)

Again, the closer to the production process, the more structured the standard work. Because leader standard work is derived from the standard work that defines the production process, leaders' standard work will vary depending on the nature of each leader's area. For example, the specifics of what a leader in a fabrication area focuses on and tracks will differ from the specifics in an assembly area. In the same way, what leaders track in an office, technical, or professional area will differ from one another, and from a manufacturing production area. In continuous process operations, leader standard work might cover reviewing adherence to procedures at critical locations throughout the leader's area of responsibility in the plant where operators have responsibilities to interact in some way with the process.

Team Leaders

Team leaders are the first line of defense for the integrity of the Lean production process. They should be on the floor in their area paying attention to their processes virtually all of the time. As such, their standard work typically accounts for most of their time—80 percent plus of their day. Teamleader standard work includes many items sequenced to the start and end of production, and many items periodically repeated to monitor and maintain the production process. Their standard work also includes discretionary time to respond to abnormalities, working on daily improvement tasks, and performing periodic tasks such as training operators. To accommodate this variability in their work, their time is specifically not filled to 100 percent.

In organizations pursuing Lean initiatives but without team leaders, I suggest considering investigating the reality of how frontline value-adding tasks are actually being performed. In my experience, it is rare for leaders removed even one or two levels from the front line to know how the work actually gets done and what it takes to accomplish it. Almost always, many opportunities for process improvement lie hidden in plain view, lacking only the added eyes on the floor of team leaders following their standard work to see and exploit them. In the abstract, team leaders seem like unnecessary overhead. In the great majority of cases where team leader positions have been added, they return many times the investment involved. And, managers who were initially skeptical of the return from this investment become the most adamant supporters of team leaders in their areas.

Supervisors

Supervisors' standard work accounts for about half of their time. Most of the items in supervisors' standard work repeat daily or weekly. They involve:

- Getting the shift started and staffed appropriately
- Reviewing yesterday's production tracking documents to understand and take any further required actions to follow up misses or other issues
- Leading the daily accountability meeting for their area, and attending the daily accountability meeting for their department or manager value stream manager's area
- Reviewing team leaders' task assignments due that day and making new assignments
- Working on or supporting process improvement task assignments

Periodic tasks during the day call for the supervisor to verify execution of the team leaders' standard work, a reflection of the slightly overlapping design in leader standard work (see, for example, Table 3.1, cells A11 and B9).

Value Stream Managers

Value stream managers' standard work accounts for approximately a quarter of their time (not including regularly scheduled off-the-floor meetings). It includes leading a brief structured daily accountability meeting as part of the daily accountability process, just as for team leaders and supervisors. For more slowly cycling processes, floor time could be less than 25 percent and accountability meetings held less frequently, from alternating days to weekly. The value stream manager's standard work includes weekly gemba walks with each supervisor for teaching and inspecting the "homework," just as supervisors' standard work includes gemba walking their team leaders. The balance of the value stream manager's on-the-floor standard work calls for verifying execution of supervisors' standard work tasks (see, for example, Table 3.1 cells B9 and C11). In this way, the value stream manager maintains his or her link in the chain of support for the integrity of the production process.

Form and Format for Leader Standard Work

Leaders' standard work differs markedly from operator standard work in one important respect. Leader standard work documents should be working documents. Leaders should have their standard work with them virtually all the time, whether on a clipboard, printed on a card, in a daily planner, or on a handheld device. Leaders should note completion of the indicated tasks on standard work forms. When they are unable to complete a task in sequence, on time, or at all, they should specifically note it and record why. This is exactly equivalent to, and as important as, noting reasons for misses on production tracking charts. Leaders' notes should also reflect when misses occurred in their areas and what action they took.

Leaders should use the standard work document to record daily notes, observations, requests for follow-up, and the like. If a leader keeps his or her standard work on a handheld device, he or she should print the record of the day's events at the end of the day. The leader's standard work documents also serve as a communication vehicle with the leader's supervisor (hence printing from the handheld). The leader turns in each day's document to his or her supervisor, who quickly reviews it to pick up any actions requested by the subordinate and note the nature of the subordinate's response to abnormalities in his or her areas. Often, turning in standard work means filing it in a visual display. This signals completion of standard work for that day and makes the documents readily available in a single location for the supervisor to scan quickly.

As part of the weekly gemba walk, the supervisor pulls the subordinate's previous week's standard work for a brief review, looking, with the subordinate, for patterns in misses (time of day, specific task, etc.) that might reveal a systematic source of interruption, something either or both might attack depending on what it is. Gemba walking (detailed in Chapter 7) is, first and foremost, a teaching and learning model; it is the main ongoing teaching method to learn Lean. The learning model implicit in gemba walking is a master-apprentice model. In this model, the master shows the apprentice how, gives the apprentice the opportunity to practice, observes the result, and gives feedback, often critique mixed with encouragement. Because Lean assumes continuous improvement and continuous learning, gemba walks are a regular ongoing activity.

For example, if meetings off the floor cause missed items in the subordinate's standard work, the superior might have to intervene with those calling the meetings. If the interruption is to attend to defective equipment, the subordinate might get a task assignment to determine the cause of the downtime and initiate or recommend action to prevent it. Similarly, if an interruption was caused by a delay in a healthcare procedure area or a holdup in discharging inpatients on schedule, the subordinate might get an assignment to investigate or recommend an action to reduce or prevent recurrence of the problem.

Periodically during this review of the previous week's standard work documents, supervisor and subordinate should consider whether the content of standard work should be updated to reflect changes in the production process or lessons learned in the preceding period. As with operator standard work, do not think of leader standard work as static. As things change in the process and as individuals learn and develop, those changes should be integrated in standard work. Standard work is not like the virtually unvarying set-it-and-forget-it atomic clock at the U.S. Naval Observatory. Standard work for operators and leaders alike is merely the best we know how to do things *for now*.

Leader Standard Work: Compliance or Improvement?

Lean is fundamentally an improvement system based on finding waste, often appearing as missed schedules, interruptions, delays, or defects. The improvement imperative is first to understand the cause, and then eliminate it. The Lean approach calls for everyone to be responsible for running their part of the business (my workstation, assigned role, replenishment route, leader standard work, and so on) and improving it.

In this approach, every standard is a hypothesis, namely, that this way of doing the job is the best way we know, for now. That hypothesis is tested every time the job is performed. Everyone's responsibility is to be on the lookout, alert to ways the job could be improved. It could be because of a problem, miss, interruption, delay, or because the person doing the job thought of a way it could be done better. The idea for improvement can be tested. If it proves actually to improve the process, you update the standard work to incorporate the improvement. That spreads the improvement to all who perform that task, a permanent (until the next one) process improvement.

So, leader standard work should include openness to and active encouragement of ideas for improving the current process (e.g., asking for or making suggestions to improve the quality of the work, make it easier, safer, more efficient, more interesting, and so on (see Table 3.1, cells A11, B9, C11, and D9). In addition, leader standard work also includes leading the accountability process to drive cause analysis of process misses and develop ways to eliminate them. By doing so, you make a permanent (until the next one) process improvement.

You might say: "Well, of course! Lean is all about improvement." But, I have encountered cases where the emphasis on leader standard work did not include this orientation to improvement. Instead, the emphasis was on compliance (Case Study 3.6 and Case Study 3.7).

CASE STUDY 3.6: LEADER STANDARD WORK—COMPLIANCE OR IMPROVEMENT?

An industrial equipment manufacturer's process had been out of control, threatening its reputation for product quality and safety. The company met production schedules through workarounds, firefighting, and heroics—at the expense of its operating costs. A domestic transplant operation of a foreign competitor presented a growing competitive threat.

The manufacturer began a Lean management initiative to resuscitate a Lean production program that had failed to sustain its gains. Its goal for Lean management was to get its processes under control. After two years, the company's operations executives were breathing easier. Compliance with standardized work and standard operating procedures had stabilized. Its operations were performing reliably. Three years into their Lean management implementation, the company's plants had many daily visual production status boards and consistent use of leader standard work. Supervisors, managers, and value stream managers all carried leader standard work with them. (One red flag, however: leader standard work documents were still the first edition from three years earlier.) Daily meetings were held at each of the production boards. But there was almost *no evidence of improvements completed or in progress*, things that should have been driven by the focus on process reinforced by leader standard work.

The visual boards were well intended, carefully put together, and neat in appearance. But, they did not call attention to problems, or actions to resolve them. Yet, a careful look at the charts on most boards suggested occurrences of delays, process misses, and defects, an impression verified by walking the floor and talking with leaders and operators. There were many relatively minor (and some not so minor) glitches, such as defective parts, missed deliveries, parts mislabeled or labels missing altogether, and processing defects. In other words, there were many opportunities for improvement, all already identified in some way. None of this was surprising in this multifaceted manufacturing process producing a family of complex products. The operations were still relatively early on their Lean journey, but running largely in a predictable state, meeting cost targets and shipping on time.

It was clear leader standard work's initial emphasis on compliance with standards and procedures had been effective. The operations were in control. In that respect, leader standard work had worked. But further improvement for this organization would require a shift in emphasis, from reporting compliance with the status quo to highlighting problems and eliminating them at root cause. There was opportunity in encouraging operators to record problems, not just the "numbers," and for leaders to shift emphasis from reporting daily status to focusing on problems as opportunities to improve. Yes, the intent of leader standard work is to guarantee the integrity of the standardized Lean production process. But, integrity in a standardized Lean process derives in part from its emphasis on the ongoing changes that come from continuous improvement.

CASE STUDY 3.7: IMPROVEMENT IS EVERYONE'S JOB

In a second case, a plant had been successfully engaged in a Lean initiative for ten years, benefitting from many improvements, most the result of large projects initiated by the plant's Lean team. The current Lean team leader had been in his position for the previous five years. He was rightly proud of what the team, under his leadership, had accomplished for the plant, perhaps a bit too proud. Detail oriented, as are many Leansters, he seemed to have become convinced that his command of the details trumped anyone else's perspective on how things might be improved. Under him, leader standard work had become a mechanism for control. For example, the leader standard work of team leaders, those closest to the experience of production operators, was printed on a laminated, pocket-sized card. The card had check boxes for when, where, and what team leaders' tasks were to be performed. You might expect that format for at least some of team leader standard work. But here, team leaders were expected only to check the boxes with a grease pencil. At the end of the shift, the cards were wiped clean for reuse.

No matter how motivated, a team leader would not be able to record notes about suggestions, problems, or opportunities in grease pencil on an 8 by 3½ inch laminated card. It was clear from the format and medium of the cards, and from conversation with the Lean team leader, that the intent was for team leaders' compliance with a set routine, rather than recording observations as seeds for potential improvements.

Meanwhile, questions about cost performance in that plant led a new operations executive to commission an independent from elsewhere in the company Lean team to do a value stream analysis of a product line's production process, an earlier star project of the plant Lean leader. The team found striking discrepancies between the actual way the process was staffed and operated and the official standard work and operator balance charts by which it supposedly ran. The official standard work and balance chart called for a dozen operators, with five of them in an assembly cell.

The emphasis on check-the-box compliance in this plant's leader standard work apparently had masked many problems and the work-arounds made by the product's frontline floor leaders. The value stream analysis team found an extra operator in the assembly cell, a "floater" who moved from bottleneck to bottleneck. Official standard work documents hung in the cell, but did not reflect any operator's routine. Perhaps most surprisingly, all this occurred in a product with a history of steady demand. The findings of the independent value stream assessment led to a new layout and rebalanced standardized work. Lead time and cost of production for the product were both reduced. The Lean team leader was moved to another position in a different plant.

In the industrial products manufacturer, an initial emphasis in leader standard work on compliance had the effect of causing leaders to over-look mostly minor but plentiful opportunities to improve. Opportunities to improve went unseen, and its operation's performance, although initially making important improvements, was now stalled and at a plateau. As for the overly controlling plant Lean leader, he had lost sight of a basic element in Lean: improvement is a responsibility of everyone in an operation, and everyone is a potential resource for improvement. The job of Lean leaders is to teach and encourage improvement by all, not to own it exclusively.

The Role of Training for Lean Implementation

It is not unusual in presentations about Lean implementation initiatives to hear how important a Lean training program has been, what was covered, how long it lasted, and how many people went through it. It is also not unusual to hear in the same kinds of presentations how much time and effort was spent in Lean training and what a waste it turned out to be, that back on the job nobody knew how the training translated into what they were supposed to do. Or, as is too often the case, superiors had not been included in the training program and so were still asking their subordinates for the same old things.

Without a doubt, there are conditions under which Lean training can be quite effective as an ingredient in a Lean conversion. It can help to drive change, develop buy-in, and move the conversion in the hoped for direction. Often, one of the expressed objectives is to help drive a change in culture. I think this misses the point about what drives cultural change, namely, expectations for different behavior and a process for supporting it.

If I had to choose between Lean training for a new recruit to lead a Lean area or providing him or her with a copy of clearly written standard work, I would choose standard work every time. Especially starting out, Lean is more about what you *do* than what you *know*, and the latter (knowing) grows from the former (doing). Training is not bad in and of itself, nor should it be avoided. On the contrary, developing familiarity with the principles of Lean through classes or reading assignments is extremely valuable. But, as far as preparing anyone to step into a Lean area and keep it operating and improving, training by itself provides incomplete preparation. Compared with specific expectations, consistent follow-up on them, and coaching in the work area from one with Lean implementation experience, conventional classroom or online training is inadequate, expensive, and time-consuming.

Training pays off far more when delivered in gemba walks through the master-apprentice learning model. Training delivered in this fashion means the lessons can be individually tailored to the student's level.

The teaching can be illustrated with specific situations in the student's area of responsibility, reinforced through assignments for hands-on application implemented over the course of a week, and followed up in the next week's gemba walk by inspection and critique. The lessons are further reinforced by the "what to do when" practices called out in leader standard work to maintain and improve Lean production and Lean management.

Summary: Leader Standard Work Is Element 1 of Lean Management

Leader standard work provides the greatest leverage in the Lean management system because it captures the expectations for executing the principal elements of Lean management. In the best of circumstances, leader standard work eliminates guesswork for floor managers and team leaders. It is typical for standard work to stabilize the leader's day. Standard work not only specifies what the leader should do, but also, by implication, identifies what the leader should not be doing. Focus on your own work and call on others to focus on theirs. The alternative is to climb back onto the fire truck. It may be a gratifying and comfortable place to be, but it drives Lean backwards. Many leaders in production operations are oriented toward doing, e.g., identifying opportunities, getting things done, crossing things off a to-do list. Leader standard work fits well with this orientation. For those with a more creative bent, standard work allows them to get the routine things taken care of with less mental energy, leaving them free to focus on making changes and improvement.

Lean management does not come about solely because its pieces are in place. The Lean management system works for you when you work for it. That means coming to think differently about what might appear to be many small things, but which add up to a big thing. The big thing is discipline on your part as a leader to follow your standard work faithfully and in depth. This entails scrutinizing the entries on visual controls, focusing on gaps revealed there between expected versus actual, and holding people accountable to complete daily improvement task assignments to address and close these gaps. All of this—closing the loop on process focus—follows from your standard work as a leader. Fueled by your disciplined adherence to it, leader standard work becomes a powerful engine for the Lean management system.

For reference, see the Lean management standards and gemba worksheet for leader standard work in Appendices A, B, and C.

Study Questions

1. How clear—and consistent from one to the other—are expectations for frontline leaders? What would you say is the effect on overall performance of the organization?
2. To what degree do mid-level leaders pay attention to the way frontline leaders do their jobs? Again, what's the effect on overall performance?
3. How would you rate the focus on the value-adding processes here? Do you think it has an impact on overall performance?
4. Do relatively minor interruptions and delays get follow-up attention to understand the causes and resolve and prevent their reoccurrence?
5. Are frontline leaders able to focus most of their time on core supervisory responsibilities related to monitoring and supporting production and people in their areas?
6. How would you say leader standard work is thought about at your workplace: as a tool to pick up and use because it seems appealing or as an element in a system that is important but not freestanding?
7. What, if anything, surprised you in reading this chapter?

Chapter 4

Visual Controls

The status of virtually every process should be visible in Lean management. If takt time is the heart of Lean production, visual controls and the processes surrounding them represent the nervous system in Lean management. Visual controls are even more important in office processes; work in offices moves in the IT network where it cannot be seen. Chapter 2 introduced visual controls as the second principal element in Lean management; this chapter offers examples of visual controls for a variety of processes. Table 4.1 lists all the visual controls that are included in this chapter and throughout the book. In addition, several photographs show what several actual in-use examples of these visual controls look like.

The intent is not to present an exhaustive survey of visual controls; instead, the purpose is to illustrate that the variety and type of visuals are as broad as the variety of production processes. The form of the visuals is limited only by your imagination, guided solely by the purpose of making easy and widely accessible the comparison of actual versus expected performance. That is the reason this book doesn't include a CD of visual control forms. The best forms are those you develop and revise yourself to show the information you need to quickly see the status of your processes.

Finally, the chapter concludes with a description of the benefits of using simple visual controls over more sophisticated IT technology.

Table 4.1   List of Illustrated Visual Controls

Figure	Title
4.1	Production pitch tracking chart
4.2	Monthly pitch log
4.3	Job-by-job tracking chart
4.4	Case-by-case tracking for variable duration cardiac procedures
4.5	Priority board color coding
4.6	Priority board hourly status chart
4.7	Completion heijunka
4.8	Late load log
4.9	Discharge plan performance tracking chart
4.10a	Detail illustration of tracking board for "re" processes
4.10b	Photo of reorder board
4.11a	Example of visual control board for noncyclical processes
4.11b	Photo of a visual control board for 5S tasks
5.1	Daily accountability board
5.1a	Photo of a daily task accountability board
5.1b	Photo of a daily run-the-business board
6.1	Visual control for progression of work through an office process
10.1a	A3 project plan form
10.1b	Photo of an A3 project plan board
11.1	Attendance matrix
11.2a	Labor and rotation plan
11.2b	Photo of a labor planning board
11.3	Sample skills matrix entries
11.4a	Suggestion system idea board
11.4b	Photo of an idea board

Visual Controls Focus on Process and Actual Performance

The purpose for visual controls in Lean management is to focus on the process and make it easy to compare expected versus actual performance. Lean is an improvement system. These comparisons highlight when the process is not performing as expected, and thus where improvement might be needed.

Comparing expected versus actual performance is a central theme in Lean management's emphasis on process focus. In some Lean conversions with an exclusively technical focus, visual controls such as performance tracking charts are often little more than wallpaper, as in Case Study 4.1. Indeed, great attention may have been devoted to developing a cosmetically consistent look for forms and displays, missing the point that Lean management is not about the consistency of the forms or the appearance of the visual tools. It is important that leaders understand *why* they track performance and that they commit to action in response to performance data and follow-through, so that action assignments turn into improvements. It may take a while before this sinks in with leaders, though sometimes it happens quickly.

CASE STUDY 4.1: USING VISUAL CONTROLS TO IMPROVE PERFORMANCE

In one company, visual display boards for cells were put up to satisfy the dictate of the division general manager (GM). The general manager insisted that the information on the boards be kept up to date. When he was in the plant, which was a half-day drive from the majority of his responsibilities, he inspected the boards carefully to make sure they were current.

At first, the boards were current only during his visits and allowed to lapse as soon as the GM left the plant. The division's Lean *sensei*, seeing this during one of his trips to the plant, asked a value stream manager to try actually using the boards for a few weeks to test the proposition that he might find them useful. During a subsequent gemba walk with the sensei, the value stream manager exclaimed that by simply noting misses and making the visual assignments (part of the daily accountability process) to respond to them, things had actually improved. Several problems of long standing were eliminated. Performance and results had stabilized. Visual controls are important, not because they satisfy executives' demands for visual displays, but because they bring focus to the process and, in doing so, drive improvements.

A Variety of Tools to Visually Monitor Processes

Performance tracking charts, such as hour-by-hour production tracking charts, are among the most commonly seen visual process monitoring tools in Lean production areas. When visual monitoring tools are implemented as part of a process that includes mechanisms to sustain them, such as leader standard work, the tracking tools have a good chance of being used effectively. Visual monitoring tools are part of a new way of managing an operation. Most production leaders are pragmatic people. If a tool works for them, they are likely to use it. But without a process that defines how the tools are used and sustains their use, the tools are likely to fall by the wayside.

Hour-by-Hour Production Tracking Charts

These basic tracking tools measure expected versus actual output hour by hour (or more frequently) during the day (see Figure 4.1). These charts are appropriate in areas with the expectation for a steady, takt-paced rate of output, such as assembly or subassembly areas in which flow production has been implemented.

Figure 4.1 shows the expected production in number of units each hour. Entries record the actual number produced and the nature and reasons for interruptions in the process. Even before a rate can be calculated, for example, in new product production areas, interruptions in flow can be documented every hour (or other interval). This is important for establishing process discipline and the habit of documenting abnormalities in the process, as well as addressing the interrupters. Reasons for misses are the most important entries on the form, regardless of the maturity of the process.

As the process stabilizes, the interval of observation should shrink from an hour to a shorter interval, perhaps a half or quarter hour, or an interval defined by a pitch. The interval can be as brief as five minutes or even less, depending on the process, its maturity, and the way output is packed and shipped. The reason for reducing the interval of observation is to provide a closer, more finely grained picture of flow interrupters. When a process is newly established, it is not unusual to experience breakdowns upstream of the process (e.g., equipment failures no longer covered up by inventory, supplier problems, scheduling snafus), not to mention problems internal to the process that cause failures. As these gross interruptions are resolved, the observation interval should progressively shrink to more precisely capture the next levels of interruption and so define the next focal points for improvement. In other words, to catch smaller gremlins, use a net with finer mesh!

Area: B211 Assembly TL: Tina T.		*Production Tracking Chart*		Date: 4/27/10 Takt: 60 sec.
Pitch	*Goal Pitch/ Cumulative*	*Actual Pitch/ Cumulative*	*Variation Pitch/ Cumulative*	*Reason for Misses*
7–7:30	20/20	18/18	−2/−2	10 min. startup mtg. Meeting long 2 minutes–safety issue
7:30–8	30/50	30/48	0/−2	
8–8:30	30/80	30/78	0/ −2	
8:30–9	30/110	32/80	+2/0	TL helped at station 5 for 6 cycles to catch up before break
9–9:30	20/130	20/130	0/0	10 min. break
9:30–10	30/160	30/160	0/0	
10:30–11	30/190	27/187	−3/−3	Container short three P/N 46230721-notified PIC
11:30–12	/190	/187		30 min. lunch
12–12:30	30/220	30/217	0/ −3	
12:30–1	30/250	30/247	0/ −3	
1–1:30	30/280	30/277	0/ −3	
1:30–2	20/300	20/297	0/ −3	10 min. break
2–2:30	30/330	30/327	0/ −3	
2:30–3	30/360	30/357	0/ −3	
3–3:30	20/380	21/378	+1/−2	10 min. cleanup washup TL helped sta 5, 3 cycles–want on-time finish
3:30–4		2/380	+2/0	Overtime: Minutes, why? 2 min., made up for part shortage @ 10:30 pitch
Totals	380/380		0/0	Pretty good shift–external failure and recovered–minimal OT

Note: Color codes used to indicate at, below, or above goal. In this black and white example, white background represents green for on goal, black represents red for below goal, and gray represents blue for above goal.

Figure 4.1   Production pitch–tracking chart.

How Visual Controls Enforce Discipline

Case Study 4.2 shows the connection between the leader's discipline and the effective use of visual controls. Visuals are an important enabler for disciplined focus on and adherence to Lean processes. This focus on process is absolutely essential for establishing and maintaining a Lean management system. And, carefully designed Lean processes require this kind of

CASE STUDY 4.2: PROBLEMS LINGER WHEN YOU TRACK BUT DON'T ACT

In a Lean conversion project early in a company's Lean journey, I found hour-by-hour charts tossed onto a bottom shelf of a supervisor's desk on the production floor. The charts were literally covered with dust. No reasons for misses were noted on them (though plenty of misses appeared); no action had resulted from them. Not coincidentally, this particular Lean conversion project stalled and did not get restarted until the product line was transferred to a different plant.

Despite early disappointments, other transformation projects were begun. As implementation and understanding of Lean management have taken hold, the same charts are posted and current, highlighted color codes readily distinguish hours where goals were met from those that were missed, while well-documented reasons for misses are reflected in posted "Top 3" weekly lists of interrupters and actions to eliminate their causes. The production tracking charts are now scrutinized daily by the area supervisor and later by the value stream manager. Periods of missed production are treated as a big deal rather than with a shrug. Support groups and line leaders are mobilized to respond to them. Overall, there is a much heightened sense of accountability for action to understand the causes for misses and to resolve them.

disciplined attention and support. We have ample evidence that Lean processes do not sustain or improve themselves. That is the reason for Lean management's emphasis on leadership discipline and follow-up, so you do not leave Lean processes on their own to fend for themselves.

Visual controls amount only to wallpaper without the discipline to insist they are taken seriously and used as a basis for action. Without disciplined follow-up by leaders, visuals are destined to take their dust-covered places behind equipment and under staircases along with the other boards and banners of past failed programs. It certainly is not a pretty picture, but it is just as certainly a common one in North American factories where leaders did not have the discipline to stay with programs they previously launched and subsequently abandoned.

Case Study 4.3 showed one way to use pitch tracking charts. The same example—instances where cells could frequently beat their pitch goals—would show up more clearly when pitch hits (green) and misses (red) were

CASE STUDY 4.3: TAKING IMPROVEMENT TO A NEW, MORE FOCUSED LEVEL

On a gemba walk with a value stream manager, I noted from the production tracking charts that two of the cells in her area reliably met and often exceeded the hourly goals. In talking with her about possible opportunities to improve in these cells, we discussed shortening the interval of observation from an hour at a time to half an hour at a time. This seemed especially appropriate in this instance.

Both cells produced a daily varying, made-to-order mix of models with a variety of work elements and cycle time. Units with less work content could be produced faster than average; those with more content took longer than the average time. If a series of lower-work-content units followed an interruption, time made up on them could cover interruptions that occurred earlier in the sequence. (The sequence was physically set in advance, so there was no opportunity to "game the system" to make results look better.)

The value stream manager decided to take that step. Her thinking was that by reducing the interval by half—in effect, looking more closely at the processes—problems would show up as interruptions that in a longer interval of observation were hidden by the cells having enough time to catch up and still meet their hourly goals.

With the smaller interval of observation, more missed pitches did show up. Interestingly, many were the result of database inaccuracies. That is, the database showed cycle times for the different units that were accurate for some models, too low for some, and too high for others. This meant the capacity for these two cells was randomly misstated depending on the units for any given day's production. This was reflected in the weekly trend chart for pitch attainment. Before the change in observation period, the percentage of pitch attainment was regularly in the high 80 percents to mid-90 percents. After the change in interval, these percentages dropped by some 20 points.

The value stream manager, with the supervisor, carefully looked at the daily tracking charts for the week after the measurement change. The supervisor pointed out the problem in the database and backed it up with documentation, identifying units that clearly ran faster or slower, showing virtually the same work content allowance in the database. The value stream manager then looked again at the tracking charts.

She asked why the "reasons for misses" column was blank for pitches where the cells had beaten or missed their goals as an artifact of the database problems. Rather than living with an inaccurate database that misstated the actual capacity of the two cells, the value stream manager wanted to see the model numbers for units with inaccurate run times recorded in the "reasons for misses" column. That would allow her to assign the value stream engineers the task of correcting the database model by model, a little at a time, rather than in a single overwhelming batch request.

The supervisor was unhappy that her performance numbers had arbitrarily deteriorated, even though previously they had been arbitrarily inflated. But the value stream manager had more confidence about being able to drive further improvement in these cells. Now that she knew the capacity of these two cells more accurately, she and the supervisor had an improved basis on which to compare expected versus actual performance. Because the value stream manager understood why she was having process performance data visually tracked, she saw the need to increase the level of resolution of the observations to provide a clearer picture of the true operation of the processes. This step led to a more accurate picture of variability in the cell and set the stage for changing from a time-based pitch to a load-by-load method of tracking. This method is more appropriate for the variable work content jobs in these cells and has proved helpful in identifying interruptions and imbalances in them.

displayed for more than a single day. It is often useful and a good practice to record the summary data in a format that makes it possible to see trends. Are Fridays more productive than Mondays, for example? Or does productivity dip after break times? The same display also allows operators in the area to see how they are doing over a period of time: e.g., "Are we getting better?" It is a simple matter to produce such a chart.

When the team leader is filling out the pitch chart with a highlighter in hand, she can also color-code the pitch on a 31-day summary pitch performance chart, one that stays in the area all month. The daily pitch charts are typically taken down and moved to the location for the daily value stream accountability meeting and kept there. The summary chart remains a local record of performance (Figure 4.2).

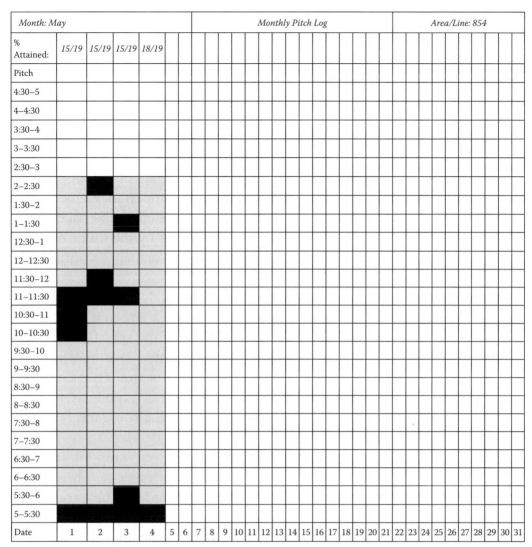

Figure 4.2 Monthly pitch log.

Job-by-Job Tracking Charts

"Flow where you can, pull where you can't, never push" is a maxim in Lean production. In Lean management, the task of visually tracking production in pull areas differs quite a bit from the expected even rate of operation in flow areas. That is, processes operating in pull (or kanban) replenishment areas are typically shared or constrained resources that supply supermarkets.

The supermarkets hold specified buffer inventories of work-in-process materials or components, each held in its own designated location, waiting for the next downstream customer processes to withdraw it for use.

The rate of resupply or replenishment depends on several factors. Among them is the rate of consumption by downstream processes and ultimately by the takt-paced final segment (often assembly) of the production process. Lot sizes and often setup times vary from job to job and part to part for single work centers that produce several or many different parts. In short, there is no even flow to measure against an expected even rate of output per hour (or other period of observation). Yet pace is an important element in Lean production. How does the Lean management system make pace visible in pull production areas? That is, in an area producing by the lot or batch instead of a piece at a time, how can expected versus actual performance be readily seen and responded to?

The answer is surprisingly straightforward, fortunately. When pull systems are set up, it is necessary to know the batch or lot size and the time allowed both to set up the equipment and to produce the specified quantity for every part produced by each machine or work center. That information is the basis for establishing machine balance charts and setting up the supermarkets and replenishment cycles on which pull or kanban systems are based. It is easy to see that this information should also be the basis for expected versus actual for these work centers.

A chart (Figure 4.3) that captures expected and actual setup and run times for each job produced during a shift in a work center clearly shows when the work is on pace and when the pace is impeded. The expected setup and run times for each job should be printed on the production instruction signal. The signal is usually a card, but sometimes might be a dedicated rack or other specialized container. Having this information on the production instruction signal makes the expected times readily available for the operator to record on the job-by-job chart. On this chart, as on most others, the reasons for misses are the most important information on it (see Case Study 4.4). The logic of job-by-job tracking to uncover delays and interruptions applies in any process where variability in work content occurs, often from one job, customer service call, or clinical case to the next. This is illustrated in a healthcare setting in Case Study 4.5.

The cardiac lab case (Case Study 4.5) is an unusual one in that the Lean intervention actually started with Lean management, the case-by-case visual control tracking chart (Figure 4.4), and the accompanying daily accountability process (see Figure 5.1 in Chapter 5.). The internal Lean

Date: _____					Job-by-Job Tracking
Operator: _____					
Asset: _____					

| Part Number | Setup Minutes | | Run Minutes | | Reason for Misses |
	Goal	Actual	Goal	Actual	

Setup and run time goals from production instruction card.

Met goal = green, longer than goal = red.

Figure 4.3 Job-by-job tracking chart.

consultant needed to act quickly to build credibility, personally and for Lean. Effectively addressing the problems of her client group, as she did, is always a good way to do that.

Helped by their internal Lean resource, the EP lab's RNs and RTs (as well as the charge nurse, nurse manager, and administrator) backed into learning Lean by documenting recurring problems and organizing for action to resolve them. For example, they learned to recognize and eliminate the wastes of waiting to start a procedure and rework in having the (already anxious, often partly sedated) patient redo the mark he or she made on his or her chest the night before.

As importantly, the team highlighted the impact on performance, capacity, and patient's experience of what previously had been experienced across

CASE STUDY 4.4: MAKE SURE YOUR EMPLOYEES KNOW THE CHARTS ARE NOT MICROMANAGING THEM

Consider the case of machine operators who produced job by job in defined lot quantities to replenish parts kept on display in a supermarket. They complained to their team leader and supervisor about being micromanaged when these charts were introduced. They felt the company was saying they could no longer be trusted to put in an honest day's work. The supervisor responded by telling them the charts were, in effect, a way for the operators to assign jobs to him! He went on to explain that his job was to resolve problems that interfered with the operators' ability to run smoothly all day. The charts would tell the story of where the supervisor needed to step in and get the problems solved. In short, he said, this was micromanagement of the production *process*, not of the people in it.

The operators began to use the tracking charts to document long-standing problems that repeatedly caused them downtime and frustration. This may well have been to test the supervisor's word that he would respond to issues the operators documented on the charts. The supervisor was true to his word. He and the area team leader responded by focusing on these problems and getting them fixed. The outcome was progressive elimination of many recurring frustrating situations that interfered with operators being able get into a rhythm and feel they had had a productive day at work.

The problems surfaced by the operators were plain to see to even moderately trained eyes, but they had simply never gotten attention in the previous batch environment, because the area had been buffered by large stocks of inventory. Now that it was running to a precise supermarket pull system, with carefully calculated on-hand quantities and replenishment times, the interrupters needed to be resolved.

Many of the issues emerged with the departure of the few experienced hands, taking with them the undocumented "tribal knowledge" of how the area ran. The result was at times near chaos, featuring things like mixing storage for active and obsolete tooling, without identification, on the same racks or even jumbled together in cabinet drawers. There were no step-by-step work instructions for complex operations, resulting in much trial and error for the new operators trying to set up and run the jobs. Raw materials were delivered in a way that required time-consuming handling to get them to the point of use. Gauges on the equipment needed alignment. Stock-outs and defective out-of-spec parts were common, given these sources of instability in the process. None of these were particularly complicated problems.

With the tracking charts as the vehicle, the operators were able to shine a light on these situations. The supervisor and team leader got them taken care of permanently, winning respect for themselves and for the tracking process as well. Stock-outs and defects from this area are now rare occurrences, operators rotate through the area without a hitch, and productivity in the area has never been as good. This is a good illustration of where micromanagement of the process can produce outstanding results for all involved.

CASE STUDY 4.5:   TRACKING VARIABLE DURATION CASES IN A CARDIOLOGY PROCEDURE LAB

A newly constructed electrophysiology (EP) lab for performing complex interventional cardiology procedures at a regional hospital had never met its projected weekly capacity. This was a problem. For patients, demand for the lab's often lifesaving cardiac ablation procedures exceeded even the lab's planned capacity; critically ill people were kept on a long waiting list. Financially, operating below capacity meant missing the target for return on the hospital's multi-million-dollar investment.

Some relatively routine procedures were also performed in the lab, e.g., implanting or servicing pacemakers and other devices. But the primary procedure involved treating cardiac arrhythmia (irregularities in the rate or rhythm of the heartbeat) by analyzing the electrical signals from the heart in order to locate and neutralize the tissues causing the arrhythmia. These procedures last as long as it takes, from a few hours up to 12 hours. Though demand and designed weekly capacity were known, cases were highly variable in duration. Tracking the process against a predetermined hourly or daily pace was not feasible. There were no goals for case-by-case performance to signal problems with cases as they cycled; the only goal was weekly capacity. But improving process performance called for an approach to identify and capture interruptions and delays as the process cycled case by case, no matter the duration.

An internal Lean consultant had worked with the cardiology areas for two years with no progress. It is not unusual in healthcare to encounter reluctance to experiment, even with administrative and other procedures that do not affect patient care or safety. There is understandable concern with patient safety, accreditation, and regulatory requirements. This reluctance to engage in Lean improvement experiments seemed to have been a barrier to Lean applications in the cardiac lab.

A new area administrator, charged with improving throughput in the lab, supported the efforts of the internal consultant. I encouraged her to seize the opportunity to make something happen, and quickly. We met with some of the nurses (RNs) and radiology technicians (RTs) who staffed the EP lab during the procedures, asking if they experienced anything that contributed to slowing patient throughput in the lab. They readily listed many frustrating interruptions and delays.

We suggested experimenting with a case-by-case (or patient-by-patient) tracking chart to identify the case and procedure, record the time in and out, and note any problems, interruptions, and delays that occurred with the case. Problems were coded with a red dot. Things that went well (consistent with the hospital's program to focus on positives) got coded with a green dot. We also suggested holding a brief daily accountability meeting. There, the previous day's red items triggered task assignments to specific RNs or RTs to understand and resolve what caused the reds. Task assignments were posted on a daily accountability board (described in Chapter 5) and reviewed for completion on the assignment's due date. To underscore the experimental nature of the project, I suggested making the tracking chart simple: hand-drawn, in pencil, duplicated on a copier, and refined as the lab team learned from experience using it.

In about six weeks the team identified, assigned, and eliminated over 50 sources of delay in the lab. The delays included supply stockouts, using indelible ink for patients to mark their chests the night before their procedure instead of ink that often rubbed off during the night, problems with nurses on the inpatient units coordinating timing of prepping patients for their procedures, insufficient patient transport personnel during the busiest times of the day, and incorrect infusion rates in the procedure for administering the antibiotic used before procedures could begin.

During this period the EP lab met its weekly target for the first time ever, and showed substantial improvement in average weekly performance. With their tracking chart, the EP team also documented the frequency of hospital-wide problems affecting procedure areas, including poor coordination with inpatient floor nurses, delays in patient transport, and incorrect orders for administering the antibiotic used in surgical procedures.

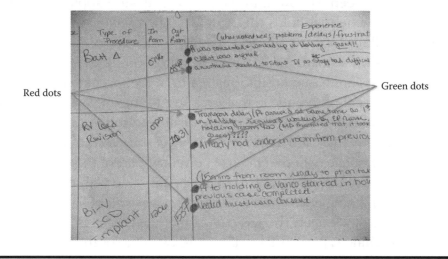

Figure 4.4 Case-by-case tracking for variable duration cardiac procedures.

the hospital as aggravations to be worked around, "just the way it is around here." To its credit, the hospital supported broader-scale Lean projects to address coordination between inpatient floor nurses and procedure areas elsewhere in the building, the imbalance between capacity and demand for patient transport at critical periods during the day, as well as resolving the issue with the antibiotic order sets, which had also caused delays that complicated unit-to-unit coordination.

It is not unusual to hear the statement: "We don't build cars, so Lean doesn't apply here." This case is a good example of Lean's principles working in a situation far beyond the paced environment of factory production.

Priority Board Hourly Status

Priority boards are a regular part of pull system operations. Priority boards provide the schedule for operators running the equipment that replenishes what has been consumed in a pull system. Production instruction cards or equivalent signals represent the schedule. These kanbans (cards or other signals) list the planned length of time to set up and run the job, the lot size, container type, and other pertinent information for the operator for each job. When the production instruction signals come back to the producing work center, they are typically put on the priority board (or rack, track, or wire) in the sequence in which they arrived. An hourly status chart for the priority board (Figure 4.5) shows the queue of jobs waiting to run for each work center. It translates the number of jobs waiting into stoplight colors.

Color codes—green, yellow, and red are typical—are applied to segments of the physical queue that holds the kanbans. The queue could be a lane marked on the floor for carts or racks leading to a work center, or a matrix of pockets, slots on a board, or a matrix taped off on the side of a machine.

The color codes are based on information from the machine balance chart. The physical segment of the queue that is closest to "next up" is colored green. Jobs in this part of the queue will normally be able to run in plenty of time to resupply the supermarket. It may not even be necessary to run the equipment just yet when the queue is in the green. The yellow part of the queue signifies a heavier load where the jobs can be replenished in time with normal work hours, provided nothing goes wrong. Otherwise, jobs will be late or it will take overtime to avoid stock-outs. The red part of the queue means a heavy load as a percent of machine capacity that will require overtime or the last jobs will be late, unless action is taken to prevent it (see Figure 4.5).

Next up	2	3	4	5
10	9	8	7	6 Order material here
11	12	13	14	15
20	19	18	17	16

Note: Color codes used to indicate load status. In this black and white example, white background represents green for capacity OK, gray background represents yellow for capacity warning, and black background represents red for capacity action required.

Figure 4.5   Priority board color coding.

The tracking chart's entry for the hour is the color code for the last job in the queue, the one with the longest wait before its turn to run comes up. The chart provides a snapshot of priority board load conditions each hour. It is easy to tell at a glance how the area is running compared to its expected status (Figure 4.6).

When this method of displaying the status of the queue waiting at a bank of equipment was first implemented, the value stream manager was delighted. "Finally," he said, "I can tell where we are in this area without having to find

Priority Board Hourly Color Status																			Asset:				Month:							
Hour																														
5 am																														
6 am																														
7 am																														
8 am																														
9 am																														
10 am																														
11 am																														
noon																														
1 pm																														
2 pm																														
3 pm																														
4 pm																														
5 pm																														
6 pm																														
7 pm																														
8 pm																														
9 pm																														
10 pm																														
11 pm																														
midnight																														
1 am																														
2 am																														
3 am																														
4 am																														
Date	1	2	3	4	5	6	7	8	9	10	11	12	13	14	15	16	17	18	19	20	21	22	23	24	25	26	27	28	29	30	31

Color = Color code for priority board row of the last production instruction card in line.

Figure 4.6   Priority board hourly color status chart.

the supervisor and ask him how we're doing. Now, all I need to do is walk by the board and I can tell at a glance whether I can move on to the next thing or whether I need to start asking questions. It's great!" Such is the pleasure visual controls can bring by simplifying the tasks of running the business day to day!

Completion Heijunka

When jobs of varying lot sizes and expected run times go through a single work center, job-by-job tracking is the tool to use for visually representing expected versus actual pace and identifying anything that interferes with expected operation. When jobs of varying work content are produced in a cell or line, the situation is a little different.

The completion heijunka identifies when the last piece of a job or load is due to be completed. It is maintained at the end of the line or cell as a pigeonhole box or any similar array of slots or pockets with the openings labeled with clock times. When the load is introduced to the process, the expected duration of processing (from the job's total work content) gets converted to the expected due-out time for the last piece. That information gets sent to the end of the line. There, the kanban (or shop paperwork) for each load is put in the slot corresponding to its expected time of completion. When the load is complete, its card is pulled from the heijunka opening. A heijunka status tracking chart lists each load and records its completion status: green for on time, red for late. The reasons for misses are entered on the heijunka tracking chart for loads that finish late, just as they would be on any other production tracking chart (Figures 4.7 and 4.8).

When time sequencing is needed to level volume or mix, a release-pacing heijunka box can be used at the beginning of the line, with expected times for starting each job defining the time slots. A job tag or kanban for each

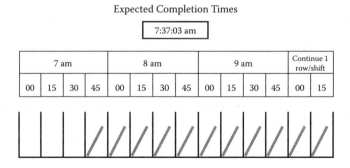

Figure 4.7 Completion heijunka.

Late Load/Job Log			
Date	Time Due	Time Done	Reason for Overdue

Figure 4.8   Late load log.

Patient ID	Room/ Bed	Discharge Time Plan	1.Nursing Assess	2.Case Manager	3.Meds Reconciled	4.Provider Assess	5.Dis- charge Orders	6.Post- care Instructs	7.Discharge Summary	Delays by step #
		Plan / Actual	Time / Initials	Time / Initials	Time / Initials	Time / Initials	Time / Initials	Time / Initials	Time / Initials	

Unit_____ Discharge Plan for ___/___/20__ Nurse Manager_____

Figure 4.9 Discharge plan performance tracking chart.

job is loaded into the slot corresponding to the load start and end times. In both cases, in addition to the heijunka tracking charts, glancing at the heijunka box and the clock tells whether you are currently meeting the expected pace.

The same logic can be used to plan and track on-time performance for discharging patients from a hospital inpatient unit (Figure 4.9). Timely execution of a discharge plan has direct implications for reimbursement in some cases. Timely discharge of inpatients is often an important factor in facilitating or impeding capacity and flow through a hospital emergency department for patients designated for admission.

Between-Process Tracking

Were subassemblies available when they were expected? Did the units come out of a separately housed process when they were due? Did parts arrive on time from the "re" processes—refinish, rework, repair, or reorder? Are suppliers delivering on time? In each of these cases, how many instances are there of overdue performance over a period of time and what are the causes? These are the usual questions you would ask from the perspective of Lean management. What is unusual is that with visual controls coded green and red, practically anybody with or without Lean experience can tell performance and performance history at a glance.

Start with an array of slots or pockets on a board labeled with clock times or days of the week or month depending on expected turnaround time. Place a tag or the actual order paperwork in the opening that corresponds to when the item is due out. If the paperwork or tag appears in a "late" spot, or if the overdue item's tag remains on the array for Tuesday at 11 a.m. and it's now Tuesday at 4 p.m., it is late.

Presto! With a glance at the board the status of performance (and reasons recorded on the accompanying log sheet) is immediately clear to all. No interpretation is required, nobody needs to make bar code scans (or maintain the bar code system) or log on to the network and look up status or a specific order, and no reports need to be run, consulted, or reconciled. Late is late, on time is on time (Case Study 4.6). The reasons are there to be seen, subjected to analysis and action right away where needed, and for root cause solution later. Note that the tracking charts for on-time performance percentages and reasons for misses are not unique to heijunka applications; they apply equally well in tracking performance between processes.

**CASE STUDY 4.6:   NETWORK DATA:
PRECISE BUT NOT ALWAYS USEFUL**

Consider these contrasting cases. In the first, units were assembled and then sent into a finishing department by conveyer through an opening in the wall and then out again for final assembly and pack, and then to shipping. Nominal turnaround time for finishing was four hours. As units entered the finish room, they were scanned in. When they were released, they were scanned out. Their length of stay in the finish room was routinely reported on a report from the IT system.

I was talking with the fabrication manager one day when we were interrupted by an urgent reorder request. It seems a unit had been in the finish room for over a week, apparently undergoing repeated attempts to rework the finish. Finally, with the ship schedule looming, finish called fabrication with the reorder. Production for that unit had to start all over again from raw materials and be expedited through the entire process. The fabrication manager, between groans, was able to tell me exactly to the minute when the unit had first gone through the wall into the finish room.

Noncyclical Process Tracking

Production is the most important process in a production operation. But without attention to its supporting processes, production will be uneven, unpredictable, and unreliable. In a Lean operating environment, production is expected to operate at an even, predictable pace, running directly or indirectly at takt time. That means many things are predictable:

- Material arrives at the point of use in a predictable manner when it is needed.
- Lab tests are completed and results turned around quickly.
- Quantities in containers and number of containers in supermarkets are as defined.
- Information from upstream departments is timely, complete, and accurate.
- Routines for handling kanban cards or other signals are accurately followed.
- Equipment is available to run when expected, operates predictably without unplanned downtime, and produces components to specification without defects.
- Rooms are cleaned and ready without delay.
- The tools, equipment, and supplies needed at any workstation are easily verified as being in their designated places unobstructed by clutter or debris.

In Case Study 4.7, the kind of tracking and the feedback it prompted throughout the process are what resulted in the dramatic improvement in turnaround time and the virtual elimination of late or split shipments due to missing units. The improvements were not difficult to make; they simply had not previously been identified and documented.

As to the computer network involved in Case Studies 4.6 and 4.7, I do not intend to make them out to be the bad guys. In the first case, we had not learned to use the information in the network. The mere presence of precise information in a database that could generate reports seemed to be all anyone was interested in. The multiple reports enabled managers in different (interdependent) departments to argue with and point fingers at each other when things were not going well. In the second case, we learned to get the information out of the network and visually display it in a way that made it far more useful for comparing expected versus actual and focusing on the process. Those steps ultimately drove significant improvement in the reorder process.

CASE STUDY 4.7: CONVERTING NETWORK DATA TO A SIMPLE VISUAL SYSTEM

An assembly supervisor was complaining to me about the lack of response to his department's request for refinished parts when the original finish was defective. The reorder requests were logged into a sophisticated software application that controlled and tracked the finishing process. The software transmitted the request to the area where parts were loaded onto the conveyor to the finish room. There, replacement parts were to be picked and put on the line or the originals reworked and then sent through the process again to be refinished. The finishing process was located, as with the first example (Case Study 4.6), on the other side of a wall and fed by a loop of overhead conveyor.

Despite the sophisticated software, orders frequently got lost somehow, usually because of human error at the ordering or load stations. The outcome was that incomplete units languished in the assembly area's so-called bone yard, the area where the kits' missing pieces were parked to wait for arrival of reordered parts. Worse, they sometimes were forgotten in the press of other work until their ship dates had passed. All involved had repeatedly been admonished to be careful when keying information into the system and to load what the system called for, but mistakes kept happening.

The supervisor never knew when to expect his reorders. Whenever he thought about it and was nearby, he would find the unload/reorder person and ask him or her, "How are we doing on refinish?" The operator would dutifully query the database. Sometimes the information was accurate, and sometimes it was not. After listening to the supervisor about this specific frustration, I asked him to think about how a visual control might help. He told me he already had a slick computer system, but it did not seem to be doing him much good. We walked around the area for a few minutes and noticed paperwork taped to racks holding kits of parts in the bone yard. The finish line software printed the paperwork for each reorder. The operators taped it to the appropriate kit rack so they would know where to put the reordered parts when they came down the line.

In a few minutes, the supervisor figured out that he could print duplicates of the reorder paperwork, on which was printed the date and time of the reorder, and place them on a board with a row for each day using adhesive-backed plastic pockets* mounted on a board at the reorder station. On our next gemba walk, he was grinning from ear to ear as he showed me the visual reorder tracking board. At a glance he could tell what parts his area was waiting for, when they had been reordered, and whether they were overdue. He no longer had to bug his unload/reorder person for status of his reordered shortages or worry about losing track of units awaiting parts in the bone yard.

Soon after that he bought a two-way radio so the person working the unload/reorder station could check with the load station on parts that were close to overdue. The original board had a row of pockets for each day of the week. If parts got delivered by the next day, that was a big improvement. In its current version, the board's array of pockets reflects an expected four-hour turnaround. The label on the pocket for reorders placed between 7:00 a.m. and 7:59 a.m., for example, shows its orders should be delivered by noon. If the reorder paperwork is still in the pocket after noon, it is easy to tell it is late (see Figure 4.10a and b).

---

* Sources for a variety of plastic pockets are Storesmart.com and Associatedbag.com.

This kind of visible tracking of performance between two processes applies as readily to processes within the same building or in the same company as it does to expected delivery times from suppliers. In many cases, the ability to distribute information across a computer network is valuable. But remember, just because information is held in databases does not mean it cannot also be simultaneously visually displayed and managed. That is always a desirable option from the point of view of the Lean management system.

Figure 4.10a and b shows examples of some of the processes that must operate effectively for an operation to run in a stable state. In Lean management, each process should have a primary visual control. Examples include a visible signal that daily maintenance and cleaning tasks or longer-interval preventive and predictive maintenance procedures were performed, and on-time completion without interruption of standardized routes through a continuous process plant or delivery routes for replenishing materials or supplies, whether in a factory, office, or healthcare facility. In addition, each of these processes should also have a secondary check or a verification that what was to be done was actually performed. Secondary checks are often items in team leaders' or supervisors'

Refinish Reorder Board

Monday

In 5:00–5:59	In 6:00–6:59	In 7:00–7:59	In 8:00–8:59	In 9:00–9:59	In 10:00–10:59	In 11:00–11:59	
Due by 11	Due by 12	Due by 1	Due by 2	Due by 3	Due by 4	Due by 5	etc.

Figure 4.10   Detailed illustration of tracking board for "re"-processes. Photo of a refinish-reorder board. On this refinish and reorder tracking board orders are expected to be returned complete in four hours. The photo is from October 27. For the day October 26, an order was put in the system between 11:00 am and 12:00 pm. A duplicate copy of the order was printed and put on the board in the slot indicating it was due back complete by 4:00 pm that day, but the board shows it is late. Two reorders are on the board for October 27. Reorder performance is visible with a glance at the clock and at the board.

standard work or items reviewed in regular weekly audits of, for example, pull system wellness, 5S, Total Productive Maintenance (TPM), or safety, to name a few.

These important chores are not glamorous, to be sure. Nevertheless, consider the saying: "An army marches on its stomach." No matter how well trained the troops are, they must be well fed over time to perform well. The same is true by analogy in Lean production. No matter how well designed the production and replenishment systems are, if they are allowed to deteriorate (a certainty if they are not looked after and maintained), the production system will run only by fits and starts, determined by who in the operation is a better hunter/gatherer.

This attention to the details of safety, maintenance, 5S, supermarket signals and levels, and replenishing materials supplies and parts is an important aspect of the Lean management system. Here, too, expected performance versus actual execution should be reflected in visual controls that act as the basis for assessment, assignments for improvement, and accountability for follow-through. As elsewhere, visual controls should be in place even if a database holds and automatically dispenses schedules. Yes, the maintenance department does have a computerized job assignment and tracking system. But how does anyone other than the maintenance dispatcher know that the work is being assigned as agreed and completed on time?

In one rather sad case, a maintenance director facing budget pressure decided to reduce maintenance staff in a plant and stretch the intervals between scheduled preventive maintenance. Because the maintenance tasks and schedules, kept in the computerized maintenance management system, were accessible only to the maintenance department, nobody was the wiser until unplanned downtime showed alarming increases and the plant manager began asking questions.

By contrast, consider the difference when the planned maintenance schedule is kept at the equipment with an easily visible sign-off spot for each date and work to be performed, plus "go see" verification written into the team leader's standard work. Compliance with the maintenance schedule can be readily verified. Proposed changes in the schedule would prompt discussion between those who depend on the equipment and those charged with keeping it reliable (see Figure 4.11a and b).

Maintaining Visual Trackers and Acting on the Information They Provide

At this point you might ask: "Aren't all these visuals a lot to maintain?" Not if there is a systematic process for maintaining them. In fact, that is one of the main contributions of standard work for leaders. Team leaders either do or do not make entries on the visual trackers as specified by their standardized work. Supervisors' and value stream leaders' standard work directs them

FF1 Support Area 5S Task Board (partial)

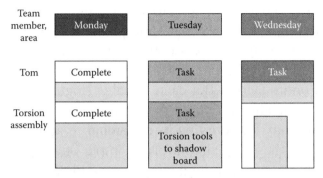

1. Task cards' "task" side color coded to match the day, e.g., Tuesday.

2. "Complete" side always green (black in this black and white figure) for at-a-glance monitoring (e.g., Monday).

3. Plastic card slots are color coded and labelled to identify task and specific day's card color (e.g., Wednesday).

4. Cards are handwritten and highlighted index cards in adhesive-backed clear plastic.

(a)

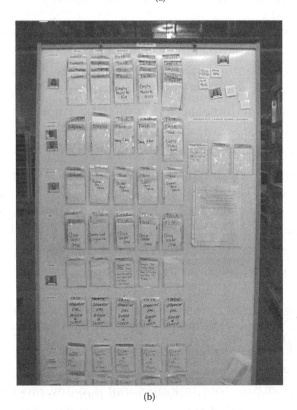

(b)

Figure 4.11   (a) Example of visual control board for noncyclical processes. (b) Photo of a visual control board for 5S tasks.

to review the visual controls several times daily (or at least once for value stream leaders) for two reasons: one is to verify the visuals are being maintained, and the second is to verify that appropriate actions have been initiated when abnormal conditions are identified on the visual controls.

When considered in total, there indeed can be lots of visual controls in a value stream or entire facility. But each visual is singular. A single person is accountable for executing it. One or two or more people have specifically designated responsibility for verifying its maintenance and for taking action if it slips. And any of these visual controls is simple, straightforward, can be interpreted at a glance, easily audited, and diagnosed. Put that together with the simple, unambiguous definition of responsibilities for maintaining and using the controls and the system looks much more manageable, as it is in actual application.

Benefits of Using Simple Visual Controls Instead of More Sophisticated Information Technology

From an IT perspective, paper visual control charts might seem like an embarrassing return to the information Stone Age. Visuals are usually not very snappy looking because they are maintained by hand. People are actually counting things (how many pieces do we expect in this load? how many are actually in it?) and writing them down. Have these people not heard of computers and bar code scanners?

It is true that many Leansters are almost reflexively anticomputer, perhaps too much so. Why? Are we simply hopeless Luddites who never learned to type or navigate in Access?

Table 4.2 lists comparisons that help explain the advantages of visual displays, and even some advantages of IT solutions. Overall, the Lean management system favors hand-completed visual controls because of its bias toward frequent eyes-on focus on the process as it cycles, and (where physically feasible) application of human assessment and judgment. Further, Lean and Lean management emphasize the importance of everyone involved in a process having timely information about how that process is performing. When information is available to only a select few, whether managers or specialists, only those few can take responsibility. Indeed, only those few have the information base for thinking about why the process performed as it did, what the causes of that performance might have been, and what might be done to eliminate the causes of interruptions or to improve performance from its current level. Consider the case (Case Study 4.8) of the high-speed

Table 4.2 Comparing Manual Visual Controls versus IT Information: Timeliness of Visual Controls

Attribute	Manually Visually Controlled	IT Provided
Info timely?	On the floor, current as of the last pitch (when maintained)	Current as of the last data entry and last time the report was run
Info accessible?	Right there for all to see, often interpretable from a distance of 10 feet or more	Right there but only for the individual right in front of the screen
Info precise?	Not always; notes sometimes vague, reporting periods occasionally missed	Yes, absolutely; regardless of accuracy
Info verifiably accurate?	Usually a close-by physical location to go see as verification of the visual display	Often can't assess accuracy; data are often a long way from the physical source and reflect human judgment and execution of data entry
Does it prompt questions?	Yes; often can be asked and answered right where the control is posted	Usually only the question designed into the report can be addressed
Easy to change and customize?	Yes; forms can be easily modified, new ones created as the need arises	Not usually; often takes IT or one with time and specialized knowledge to make changes—assuming desired changes don't crash other applications
Intimidation factor?	Very low; about as difficult as coloring with crayons	Can be quite intimidating to those not literate in the system
Ownership or alienation?	Floor operators create the information that stays in their areas, often in their own handwriting	Information is taken away from the floor, transformed into computer-generated reports that no longer have "fingerprints"
Info simultaneously available many places?	No (with exception of multiple card kanban systems)	Definitely a strength of an IT network
Computational accuracy?	With humans, no computation should be considered routine!	Definitely a strength of an IT system
Overhead required?	Very little beyond people who are already there; pens and highlighters are cheap!	Thousands and millions for gear, specialized departments, consultants, troubleshooting, etc.

CASE STUDY 4.8: HIGH-SPEED CARTON PRINTING: WHO KNOWS WHAT AND WHEN?

A high-volume printing plant produced color paperboard carton stock on two presses for consumer beverage packaging. The four-color high-speed presses, each several hundred feet long, printed and scored carton stock that was shipped flat for later folding and gluing at its customers' filling operations. The presses were fed by large multiton rolls of blank paperboard that were manually positioned and fed into the presses at the beginning of every run. Each press produced hundreds of printed flat carton stock per minute.

Press controls and performance measures were located in a second-story glassed-in control room overlooking the span of the pressroom floor. On the floor the continuous web of paperboard went by in a blur—until a sound like a deck of giant cards being violently shuffled brought the press to a stop. The paperboard web had gotten misaligned at a series of scoring die-cutting stations near the end of the process. Mangled carton stock had backed up and spewed over the top and sides of the press. It was a mess that had to be cleaned out by hand and the press rethreaded before operations could be started up again.

The few operators who worked on the floor tending the presses, when asked how frequently something like this happened, replied: "All the time." There were a few whiteboards on the floor with charts lined off and labeled, but nothing written on them. When asked where the data about the misalignment we just witnessed were recorded, an operator pointed up at the control room, explaining the machine operating data were automatically recorded "upstairs for the engineers." The operators said they were not asked to participate in debriefing the pile-ups like the one that had just happened. For their part, the engineers had not been not able to get to the root cause of the web misalignments that spewed scrap carton stock and periodically brought production to a halt. The plant manager, however, expressed great pride in the auto-populated databases documenting the plant's operating performance.

carton printing line as an illustration of this gap between information from programmable logic controllers (PLCs) and experience from the floor.

In many industrial processes the nature of production is too fast or dangerous for people to have direct hands-on participation in production. And of course it makes sense for automated collection of operational data from process assets. But even these operations have people present somewhere in the process to observe parameters, and perform maintenance tasks, clean and change filters, manually load or stock up particular stations, make rounds to watch for early signs of characteristic problems and intervene to prevent them, and so on. The PLCs and automated data recording gear typically are not set up to feel, hear, see, or smell a change in the way the equipment is operating, or to reason what conditions prompted a particular manual adjustment.

Both Visuals and Automated Tracking

This "scene of the crime" information based on human sensory data is valuable. The eighth waste in Lean calls out underutilized people. It is an example of this waste to ignore these sensory data, not ask that it be logged, and leave out of the cause analysis and problem diagnosis process those who were present when the operation started to go off track. The eye witnesses may not provide all that is needed to solve the problem, but what you asked them to observe and record is likely to provide a richer, more nuanced picture that helps you understand and interpret your autopopulated databases.

Consider the first two attributes in Table 4.2. Timely information about process is important for two reasons. First, if the information on the visual tracking is filled in, it at least means someone is in touch enough with the process to record its performance. And the information is widely accessible, particularly to those working in the process. Often (and preferably) thanks to color coding, the information is interpretable at a glance from a distance of 10 feet or more. (Chapter 5 includes a case study (Case Study 5.1) where the operators at the far end of an assembly process brought binoculars to work so they could see the production tracking chart and know if they were on or behind pace!)

With information captured by an automated process, the information will be timely as long as the automation is working and data entry has occurred on time. But the information is often not readily available to the people working in the process because computer and network access is required to view it. Few may have login privileges or the licenses required by SAP and other software suppliers. Or the computer may be located where only the person operating the system can read it. Further, if the information requires manual data entry, such as scanning production paperwork, it may be current only a few times per shift. Finally, the ability to read the all-important reasons for misses entries requires one to be right in front of the screen.

Accuracy versus Precision of Visual Information

Computerized information is highly precise, often notoriously so. Results may be reported out to several decimal places, but based on inaccurate entries or relatively old data. Especially where reason codes are the basis for recording reasons for misses, they may convey a false sense of precision. Indeed, they may convey a view of conditions from the perspective only of the person who developed the codes. Coding schemes often lose their bearing on current operations as conditions change, or are too general to convey what actually occurred, or are checked off quickly with little consideration for their specific applicability (Case Study 4.9).

Here is something to keep in mind when setting out to collect process data, or data of any kind. Data collected in discrete or disaggregated form can always be aggregated into categories. But data collected in aggregated form can never be disaggregated into discrete elements. From Case Study 4.9, if the setup and changeover problem involved the server balking at providing the right program, or there were problems loading the workpiece, or the tooling magazine was jamming, or the chuck was worn, or the setup information sheet left out information on initial distance between the cutting tool and workface, or the tooling appeared worn and the backup tool that should have been in the job kit was missing, or who knows what else, the bar on the Pareto chart is silent, saying only: "Sorry friend, it's all just setup and changeover to me: figure it out yourself!"

Instead, start out and stick with asking your frontline workers to record in their own words what happened. Here you can ask people to tell you what they had but did not want, or what they wanted but did not have. Without signing anyone up for a class, you are teaching them the first step in problem solving, which is to define the problem. As you work through the problem with them, you are modeling the other steps in problem solving, even if only to ask "why" five times. The only other alternative is to throw out the aggregate data on the Pareto chart and start over, asking people to write what happened in their own words as new setup and changeover problems occur, then looking for patterns you can meaningfully aggregate and further explore.

When hand-entered reasons for misses are reviewed, it is easy to assess the quality of information presented. Is the reason a well-crafted problem statement, such as "light paint on upper inner surface of drawer handles on series XXX models"? Or is it too vague to interpret, much less to use as

CASE STUDY 4.9: REASON CODES ARE A BAD IDEA

I was visiting a contract precision machining plant that had completed a first-pass implementation of visual controls for its operations. The shop floor consisted of rows of sophisticated CNC lathes and machining centers, most of which were leased and thus relatively new, and well maintained. Appropriately in an operation like this, the visual controls were focused on material availability, scheduling, and equipment downtime. The plant Lean leader indicated, not surprisingly, that most downtime occurred when changing over from one job or job family to the next. The new job's material and tooling had to be loaded, the new program called up on the machine, and the setup tested before the job could be started, the first piece checked, and the machine put into automated operation. Surprisingly, there were no expectations by job for how long a changeover should take, nor was there standardized work defining the sequence of steps for any given changeover.

Each machine had a daily log sheet covering both shifts. Operators recorded each job, how long it took to complete, number of pieces produced, and other information. A preprinted Pareto chart on the log sheet listed reasons for downtime. Over the course of the two-shift day, operators filled in segments of a bar chart representing the amount of downtime they experienced based on the listed reasons.

Looking at one of the log sheets, by far the longest bar, and thus the most lost time, was for "setup and changeover." I observed that it was equivalent to explaining why people were hospitalized by the category "illness and injury." It tells you everything, but in such a general way as to provide no information on which you can act. And in this case, if you did want to act, you would have to rework the Pareto entries, trying to recreate the situation at the time each incident of downtime occurred. Not only is it a waste of time, but also fallible memory is likely to produce inaccurate or incomplete information. Reason codes may seem neat and efficient. In the practice of process improvement, they are a bad idea.

a basis for action, such as "handles" or "paint?" Is a reason for interruption missing altogether? This is an instance where a second-level process check can occur, in this case of the quality of the entries on the visual controls. This kind of opportunity is rarely available in automated data collection systems, or on systems that rely on reason codes.

Proximity of Visual Controls

Another advantage of visuals completed by hand is that they usually are (and should be) close to the process whose performance they reflect. That means it is easy to go look at and verify that what is shown on the visual corresponds to physical reality in the production process itself. This ease of verification is another example of assessing quality of the information in the visual control.

Sometimes it is possible to make the same verification with IT systems, but often the relative scarcity of computers in the production area makes it less convenient to perform a go-see verification or quality check. Worse, the computerized system encourages managing the production process from a computer screen in an office somewhere removed from the actual production area. Interestingly, this problem can also occur in supervisors' offices even when these are little more than a stand-up desk and a computer. When you are focused on a screen, you are not focused on the process. Either violates the Lean management system's three-part prescription for focusing on process: "Go to the place, look at the process, talk with the people."

Flexibility of Visual Controls

IT systems provide powerful analytical tools, without doubt. However, often the analyses (that is, the questions addressed by the automated systems) are those programmed into the reports that are accessible on the system. Typically, only the questions addressed in the report can be answered by it. In contrast, questions prompted by entries on a visual control chart can often be addressed, at least initially, right where the control is posted. When that is not the case, for example, when a new defect has surfaced or supplies have been exhausted and not replenished, a new control can be easily drawn up to track and address it.

For example, when a piece of equipment starts malfunctioning, it is easy to draw up a simple downtime log to record instances, durations, and observable causes or symptoms that can be tallied up or Pareto-charted. Or, when a process starts to produce rejects, it is a simple matter to start tallying them. Pareto-charting the defects and then taking action on the findings come next, with no programming or transcription of data required. By contrast, someone with specialized knowledge is usually required to reprogram an analysis in an automated system. It might mean getting support from IT, often not a quick proposition, or getting help from a technical professional in the area, probably already working on a full plate of assignments. Or automated tracking, specifically because it is automatic, can produce nicely printed charts from, say, automated bar code scans of completed units. When actual volume is less than expected, nobody is there to record the miss and the reason why. Or it happens at the end of the day, as in: "Let's see, why did we miss that pitch seven hours ago?" Long batches of time of this sort often produce less than accurate and less than useful information.

Visual Controls and the "Fingerprint Factor"

Perhaps it should not be so, but often people in a production operation look past posted computer-generated reports, even graphic ones, of leading performance problems. That is especially true when the data are reported in table format. Managers often do not realize that it takes a trained eye to read and interpret information presented in these ways, and production operators often have not had that kind of training. As an example, I recently saw a daily quality report, an Excel spreadsheet printed in tiny type (to fit on one page), posted on a shop floor start-up meeting board. I asked if anybody used or could even read it. The answer: "No, not really." It is a very different story when the operators themselves have been involved in recording the data, or working through it in start-up meetings led by their team leader.

Hand-created data, especially when you know whose hands created it, including the real possibility that the hand was or could have been yours, have a much lower intimidation value than the crisp, precise-looking management document someone has posted on your team's information board. This fingerprint factor is important. It helps draw people to the information and conclusions from records they have had a recognizable part in creating. The same is not the case with impersonal, computer-generated materials, even those with the slickest graphics.

The Power of Networks

There are two aspects of information reporting where IT solutions win hands down. IT networks are excellent tools for broadcasting information to dispersed locations. A network can readily broadcast the sequence of units starting into the final assembly process, so that subassembly operations can produce to the discrete unit in exactly the needed sequence (Case Study 4.10). A version of this is at work every time a customer places an order at Taco Bell. No finished inventory needs to be held; instead, every order is custom-produced in the proper sequence.

CASE STUDY 4.10: USING THE NETWORK AS A CALCULATOR

Another application where IT shines is where mathematical computation is required. Computers are far more reliable at simple math than humans. Here is a case of a creative, early use of network broadcast subassembly signals similar to those discussed just above. In this case, the need was to signal the number of drawers that would be needed in final assembly of storage cabinets. The cabinets varied in size from two to five drawers and came in three widths. An early solution was to hold the drawer bodies, all of which were a standard interior color, in a supermarket on the assembly floor. The drawback to having all the sizes on display was the amount of floor space they consumed and the large number needed to guard against stock-outs. It amounted to a lot of inventory and a lot of floor space.

The broadcast solution was to send the drawer producing area the information on the number and sizes of drawers for cabinets just entering the color application process prior to assembly. Operators would call up the proper screens on the computer and do the math to determine the number of fixed lot quantities of drawers in each size that were called for. They would then put a kanban for a lot quantity of that size drawer on the priority board. In the course of the multiplication and division required, the numbers did not always come out right. Sometimes too few drawer bodies of a given size were prepared, and sometimes too many.

The solution quickly became apparent: have the computer do the math and then broadcast the number of fixed lot quantity orders to the drawer prep area, signaling the operators to put the right kanbans on the priority board. It was immediately clear that having the computer compute and broadcast was far superior to having the operators call up screens and do the calculations. The result overall was a significant reduction in inventory and floor space and much improved availability of the right parts at the right time.

This is a particular application of the golf ball method of signaling developed in Japanese factories. The difference is that instead of color-coded golf balls arriving at a subassembly station to signal the model or type of unit starting into final assembly, an electronic signal arrives on a screen or prints as a label at the beginning of the subassembly process—just like an order for a bean burrito with no cheese. These are flexible and powerful applications, as long as the server does not go down!

There are applications of kanban systems where the same information (what part, how many, which container, etc.) can be broadcast for display at multiple locations, for example, at the ordering and producing locations. But where kanban cards might have to travel a distance (surprisingly short—50 feet—for this risk to surface), there is a real danger of kanbans being lost. In these cases, broadcast transmissions of the kanban information, displaying the kanban at the sending (or ordering) end, and printing a copy of the kanban at the supplying (or receiving) end are the better solution, preserving the advantages of visual management.

Intangible Benefits of Visual Controls

It is probably beyond dispute that asking people to take a few minutes per shift to record performance data requires far less overhead than the cost of hardware, software, and support resources needed to automate data collection to track process performance. Beyond the financial implications is the "soft" benefit no accounting system can calculate. These are the benefits from increasing the level of involvement of operators in observing, analyzing, and improving the processes in which they work every day.

Overall, visual controls not only heighten focus on process and accountability for that focus, but also provide the foundation for a far greater level of employee involvement than could any other reporting system. For Lean production to truly be a process improvement system, that kind of involvement is essential.

Summary: Visual Controls and the Data for Lean Management

Visual controls are powerful contributors to Lean management. Visuals reflect the human scale of production activity and process focus. They connect people to their processes, and at the same time reflect adherence to the process—or its lack. In doing so, visuals represent an important, simplifying addition to Lean management.

Visual controls help transform the abstract concept of discipline in Lean management into directly observable, concrete practices. Actual versus expected moves from the realm of an idea to easily interpreted visual tracking tools. Visual controls are the basis for comparing actual versus expected performance in Lean management. They do not make the comparison, but they make the comparison possible, if not unavoidable. Visuals are a principal vehicle for focusing on process, and for verifying this important aspect of Lean management. The variety and application of visual controls are limited only by your imagination. Which visuals to use and where to use them are determined by the needs, both stable and emerging ones, in your processes.

The actual appearance of visual controls is relatively unimportant beyond basics of legibility and ease of interpretation. I have seen both plain and fancy ones work equally well or poorly. What is important about visual controls to you, the leader at any level, is that you understand the reason for having them. By insisting through your standard work that the visuals are maintained and current, visuals constantly reinforce the focus on process. This focus makes it easier to see the contrast between expected and actual process performance. By doing this, visuals allow you to identify opportunities to press for improvement.

At first glance, visual controls may look like an afterthought or mere window dressing to the serious business of Lean technical design. In fact, visuals provide the basis for much of the management—process focus, discipline, and accountability—in the Lean management system. And the Lean management system is necessary to sustain Lean production and extend its gains. Neither aspect of Lean, technical design or management system, stands well without the other; visual controls contribute substantially to the robustness of Lean management and, through it, Lean production.

Finally, visual controls in Lean do not always mean the absence of IT tools, but the immediate, accessible, flexible, inexpensive, and responsive nature of visual controls makes these simpler solutions the preferred ones for most applications in Lean operations.

Study Questions

1. Do your visuals mostly display status or history, or do they reflect a focus on process that highlights problems to drive improvement with visual evidence of improvement activity?

2. Can you trace back a connection from process improvements to problems identified on visual controls? That is, does your investment in visuals yield a return in implemented improvements?

3. Are visuals here only visible on computer monitors? How effective are they in driving improvement? In engaging the workforce?

4. Are your visuals modified based on your experience with them, and as processes change?
5. Who creates the visuals in your workplace? Who records information on the visuals, and how close are they to the production process? What is the impact on the visuals' effectiveness?
6. Is there an emphasis on a consistent look and format for the visuals in your workplace? If yes, how effective are they and do they work equally effectively everywhere?

Chapter 5

Daily Accountability Process

A daily accountability process is the third principal element of the Lean management system. It provides the steering wheel, directing which improvements will be worked on. The accountability meeting leader makes task assignments first to understand the cause of a problem captured on a visual control, then to eliminate the causes. The accountability process also provides the throttle for improvement, the due date, and resources for the improvement tasks. This chapter shows how to implement daily accountability through daily meetings and visual controls, with case studies to guide the way.

At first glance, this component of Lean management seems designed simply to ensure follow-up on task assignments made in response to yesterday's problems or opportunities for improvement. A more significant (though less obvious) purpose of daily accountability is to reinforce the Lean management system's focus on process and, through it, to identify and implement opportunities for improvement.

Note as well that without an accountability process, visuals are likely to be empty exercises, lacking the means to resolve problems captured on them. I have heard frontline workers who have seen various visuals come and go refer to them as "just signs and lines" that had no effect on reducing the frustrations of daily interruptions and delays on the production floor.

How Conventional Production Differs from Lean

In the conventional production world, the object is to meet the schedule. Leaders are expected to fashion workarounds to get past problems that threaten schedule completion. There is little follow-up because the task is the same the next day: do whatever it takes to meet the schedule. In a Lean world, the focus is on maintaining and improving the process. Follow-up in Lean management calls for understanding the causes of yesterday's problem and then eliminating them.

In a conventional or batch system, you ask: "Did you meet the schedule, or hit the numbers?" That is pretty much the sum total of your daily accountability. In a Lean system, you ask: "What caused the problem that interrupted the process, and who will do what to resolve it?" Where conventional systems have structured processes to deal with daily shortages and recurring delays, for example, Lean management's structured process deals with daily accountability for making process improvements that eliminate the causes of shortages.

Lean management is all about focusing on the process: stabilize it, standardize it, and improve it by exposing problems, eliminating them, and repeating the cycle over and over (Figure 5.1). Visual controls are a key in making it easier to see quickly the status of your processes and the problems that interfere with them. The design of standard work for leaders, slightly overlapping and redundant, is intended to ensure the visuals are maintained in the first place, and acted on after that. Again, this is a Lean management system for closing the loop on process focus and driving improvement.

So in addition to including regular checks on the status of visual controls, leader standard work also includes a process to follow up on the stories told by the visual controls. This is the daily accountability process. In it, leaders assess the meaning in the visuals, assign appropriate responses, and hold people accountable for completing their assigned tasks. This follow-up process occurs largely in the structure of daily, or weekly where appropriate, three-tier meetings. In this way, the three principal elements of Lean management combine to form the essence of the Lean management system. Leader standard work, visual controls, and the daily accountability process, plus the discipline by leaders to maintain the integrity of these three, form the heart of what makes Lean management go.

Three Tiers of Daily Meetings

The daily accountability process typical of a reasonably quickly cycling daily work process takes place as an interlocking set of three brief, structured, daily meetings, one of which is the familiar, but often misunderstood,

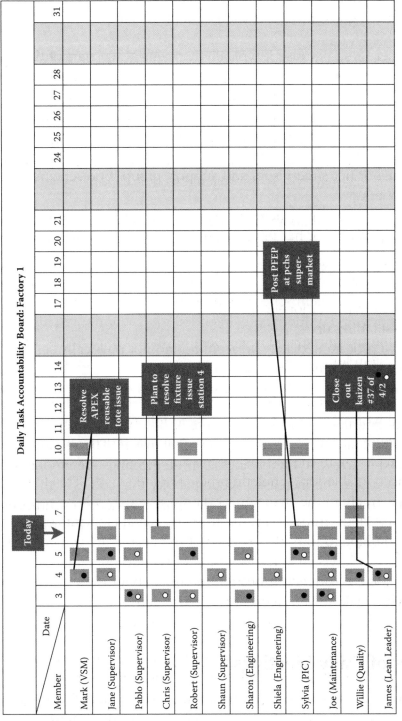

Note: Color codes are used to indicate tasks completed on time or not. In this black and white example, black stands for red, white for green.

Figure 5.1 Stabilizing and improving: Lean production and Lean management work together to produce continuous improvement.

team start-up meeting. Each of these meetings is an explicit example of Lean management's focus on comparison of expected and actual.

As the name suggests, there are three tiers of meetings:

1. The first tier (first, because it typically happens at the start of the shift or workday) is the work group team leader meeting briefly with the team members.
2. The second is the supervisor meeting with his or her team leaders and any dedicated support group representatives.
3. The third-tier meeting is the value stream manager or equivalent meeting with his or her supervisors and support group representatives or staff members.
4. A fourth tier is possible where the plant, unit, service, or building manager meets with his or her production and support staff members.

Each of the meetings shares these characteristics:

1. Brevity—rarely if ever longer than 15 minutes.
2. Posture—standing up.
3. Location—on or immediately adjacent to and not physically separated from the unit, service, or production floor.
4. Agenda and content—defined by a visual display board.

The second- and third- (and fourth where applicable) tier meetings have a dual focus. The first is on "run the business" activities. The second is on activity intended to improve the business. These meetings are among the main occasions where the increased accountability that comes with visual controls can be seen. Accountability and disciplined adherence to process are central to a daily enactment, primarily at the second- and third-tier meetings, of a three-part process:

1. Assessment based on data captured on visual controls
2. Assignment for corrective action or improvement
3. Accountability for having completed the assignments due today

Far from the daily bureaucratic burden of typical recurring staff meetings, these second- and third-tier daily stand-up sessions are highly structured with the focus on building accountability and taking action to resolve problems and drive improvement. The first-tier team leaders' meetings are, among other things, the primary vehicle for supporting and extending bottom-up participation in improvement suggestion systems. In the case of all three (or four) tiers of meeting, information and accountability for performance are clearly, graphically, and visually displayed.

The daily three-tier meeting structure itself is something of a visual control. The meetings take place on or near the production floor standing up at the information board for the appropriate tier. The board is the location, agenda, and content for the daily meetings. The agendas and roles are standardized. The meetings are brief, not exceeding 15 minutes for the third tier. The team leader, supervisor, or value stream manager leads his or her respective meeting.

The first agenda item is always the day's labor plan. At the first-tier meeting, this item is addressed as a visual display of the day's labor or staffing plan showing starting positions and other assignments for the day. Did all team members who were expected show up and check in at the team meeting? If not, who is assigned to what job to get the day or shift started? At the department meeting, if unplanned absences are interfering with the start of work in one of the areas, units, or production teams, how can we shift people among areas quickly to bring production up to the day's planned rate? Similarly at the value stream, has any department been unable to balance labor with the day's planned work? And if so, what redeployment can be made at the value stream level to get things running as planned? Beyond that, the agenda and scope expand going up the tiers, as described next.

Tier One: Team Leader and Production Crew

The focus at the production team's meetings is mostly on today's assignments and any items of special note that day. The daily rotation and labor plans for the area (described in Chapter 11) are on display at this meeting, so people can check their assignments before work begins. The team leader updates yesterday's performance, covers any items of note from the previous day, and reviews today's plans and any issues. The team board may display summary project plans (objectives, current and planned future states, indicators of success, timelines, members) for significant externally supported activities going on in the area. And, on designated days of the week, specific topics might receive special focus. These could include, for example, safety, quality, 5S audit results, status of ideas suggested by team members for improving processes in the area, or other topics.

Especially for this meeting, the principle of pull communication is important to observe. That is, it is easy to provide far more information than is relevant, or than team members are interested in or prepared to act on at any given level of system maturity (recall the quality report in the tiny eye-test type in Chapter 4). These meetings are not intended to become one-way communication conduits for the organization. They are intended to be a place where team members feel free to raise questions and concerns and where team leaders and others respond or do the investigation necessary to provide a response. Yes, some corporate news is passed along as well, but team members have an active role in putting topics on the information sharing part of the agenda.

Where there are not yet well-defined or well-understood reasons for maintaining team information boards, it is common to see the boards, even elaborately designed ones, festooned with printed copies of emails, often weeks or months old, and hung with Excel output—spreadsheets, charts, and graphs. That is usually a sign of confusion over the purpose of the board. By contrast, employees will make it clear when they are interested in information that reflects their performance. The binoculars case (Case Study 5.1) illustrates that as well as any I have seen.

Tier Two: Supervisor and Team Leaders

This meeting, led by the supervisor, focuses on two topics: running the business and improving the business. Information on the departmental team board typically includes status of key processes and equipment as well as week- or month-to-date summary performance data on safety, quality, delivery, and cost (or productivity). Team leaders bring their production tracking charts from the previous day and post them on the designated spot on the information board. The board is also likely to display the previous week's top three problems, their magnitudes, and actions underway to resolve them.

The initial items on the meeting agenda are similar to those from the team leaders' meetings: noteworthy items from yesterday and upcoming issues for today, as well as updating the trend charts on the board. The focus then shifts to the production tracking charts. It is important for the supervisor and his or her team leaders to understand each problem, interruption, or delay captured on the tracking charts. These will be the subject of further discussion at the third-tier (value stream or equivalent) meeting where the production tracking charts will be scrutinized again. The supervisor knows to be prepared to describe what happened; what actions, if any, are underway in response; and whether he or she needs support from anyone at the third-tier meeting.

CASE STUDY 5.1: DISPLAYING FINISHED GOODS BY UNITS OR DOLLAR VALUE

In a cabinet assembly area, there were two parts to the main assembly line. The cabinet cases were built up at the head of the line and then moved in first-in, first-out (FIFO) sequence into a queue between case buildup and drawer assembly, insertion, and final assembly. (In a later refinement of this layout, the drawers are assembled and inserted as the cases are built up, all in the same line.) The whole process was designed to run at a takt time of 36 seconds, producing a completed cabinet every 36 seconds. A production tracking chart was used to keep track of the line's pace every half hour. The team leader took the number of finished units from the counter at the end of the line, wrote that number in the "actual" column for the time period in question, and then picked up a highlighter and color-coded that period's actual:

- Green signified having met the takt pace.
- Yellow meant missing it by 5 percent or less.
- Red meant the line had missed the takt pace by more than 5 percent.

(In the refined design, only red and green are used.)

The drawer assemblers were separated from the head of the line by about 50 feet. The production tracking chart was beyond the head of the line, far enough away that the drawer crew could not see it. One day the value stream manager noticed a pair of binoculars in the drawer assembly area and asked about it. One of the operators told him he was unable to see the color code on the pitch chart because it was too far away, so he brought in the binoculars to be able to tell how the line was performing.

In a similar case, team members in a shipping department used to pay little attention to the posted information about the dollar value of a day's shipments. When the team leader switched to writing the average dollar value shipped per team member and a goal for average dollars shipped per member, the operators immediately began to pay attention, becoming interested in tracking their performance at several points during the day.

Few of us would participate in a competitive game without keeping score. Most people who come to work every day have some level of interest in knowing how they or their team performed. That interest is a pathway to what motivates them. It is worthwhile to reflect those interests with items on the team's tier one board.

The second topic, improving the business, involves a visual task assignment board (Figure 5.1). This is a matrix with rows, one for each meeting participant's name, and columns, dates (from one to eight weeks) labeled day by day. The supervisor makes assignments to the team leaders and any support group representatives in response to needs identified from the production tracking charts and from the other improve-the-business tools that are regular features on the meeting agenda. The supervisor writes these assignments on small cards or Post-it® notes and places them on the board on the due date for the person assigned. The supervisor's weekly gemba walks with the team leaders can result in assignments, as can less formal observations in the area. Assignments can be follow-up items from regular audits (such as 5S, supermarket status, safety). Actions assigned to the supervisor at the third-tier meeting or on gemba walks with the value stream leader might result in subassignments to team leaders or support group members in connection with the supervisor's task assignment.

The Green Dot/Red Dot Convention

At each day's meeting the supervisor reviews the tasks due that day. He or she works down the column for today's date, asking for each assignment in today's column if the task is complete. If it is, the supervisor puts a green dot with a marker or green adhesive (sticky) dot on the Post-it note. If it is incomplete, the supervisor pursues the reason enough to get a clear understanding, asks when the expected completion date will be, negotiates about that if necessary, and puts a red dot on the Post-it note. The assignment is kept on its original due date. The first assignments to be reviewed in this part of the meeting are always those that are overdue (see the dots in Figure 5.2a and 5.2b).

The original due dates for assignments are never changed, nor are the assignment Post-its ever moved from the original due date. The green dot/red dot convention is absolute and arbitrary. Complete is complete; overdue is overdue. Period. When tasks are completed, the meeting leader puts a green dot next to the red one.

Lots of red dots accumulating on daily task assignment boards invite important questions:

1. Are we overestimating our capacity to get assignments done?
2. Should the meeting leader consider adding more capacity, focusing existing capacity better, or recalibrating his or her estimates of what is entailed in a day's improvement task?
3. Are tasks being kept from completion by impediments external to the area?
4. If so, should the meeting leader or a designated other investigate?

Impediments to task completion external to the area sometimes need to be brought to the next higher tier for investigation, support, negotiation, or resolution.

Think of the red dots on task assignment boards the same way you think of reasons for misses on any visual production tracking chart. Their purpose

Figure 5.2a   Daily accountability board. Photo of daily task accountability board. This photo is from October 27, the date marked by the dark index line. The assignments from earlier in October have been removed. The board's magnetic date numbers will be reset for November when the month turns. Color coding is done here with markers. The dots are visible on incomplete assignments transferred from October to November. The color-coding markers are held on the board by magnetic tape. The names of the value stream manager, supervisors, and support group representatives appear at the left.

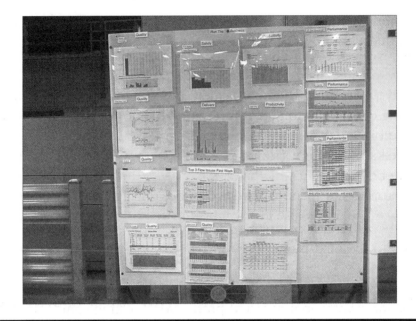

Figure 5.2b   Photo of daily run-the-business board for a focus factory. Boards like this one show yesterday's pitch attainment and other visual controls for performance appropriate to the process. They also display yesterday's entry on trend charts for safety, quality, delivery, and performance. Any of the charts can prompt process improvement assignments. These can include assignments on the daily task board, or result in a problem-solving or project assignment.

is neither shame nor blame; rather, they highlight a gap between actual and expected. In the relatively rare case, red dots will accumulate because someone is either incapable or unwilling to get his or her assignments completed. That becomes a matter for a performance discussion with the individual involved. Most of the time, something unexpected has happened or the planned cooperation of an external group or person has not come through. If the dots form a pattern over a week or more, the meeting leader will definitely want to investigate.

Stating that red dots do not represent an individual failure is easier to say than to hear. Even assigning visible, public accountability for completion dates is enough to make many uncomfortable at first. That is why it is important to emphasize the arbitrary nature of the green/red rule. Things do happen that are outside any individual's control; that is the most frequent cause of red dots. But experience with this process in many organizations around the world is that people soon come to see they are getting credit for getting tasks done on time. And, people learn that nobody makes a big deal about an occasional understandable red dot here or there. Really.

Day-to-Day Project Management

To be effective in the daily improvement segment of the three-tier meetings, the meeting leaders (supervisors and value stream leaders) need at least rudimentary project management skills (a topic touched on in Chapter 9). That is, the leaders must quickly perceive the series of step-by-step actions to attack a flow interrupter or develop an improvement. This skill, the ability to see an implicit work breakdown structure, is necessary to make appropriate one-step-at-a-time task assignments that cumulatively respond to the interruption or opportunity. Follow-up on these task assignments is straightforward with the visual daily task board on which the assignments are posted.

Tier Three: Value Stream Leader with Supervisors and Support Groups

This is the daily value stream (or equivalent) meeting at the value stream board. (Again, for less quickly cycling processes or for larger-scale project reviews, the tier three (or four) meeting might be held weekly.) At this meeting, daily performance data are added to trend charts covering the typical dimensions—safety, quality, delivery, and cost—along with any others important to the area.

The value stream manager briefly reviews the day's staffing situation and then touches on yesterday's performance measures. Items of note are covered regarding past performance or upcoming items. Next, the value stream leader turns to yesterday's performance tracking charts. He or she scrutinizes them for completeness, misses, and the reasons why. Based on the data, he or she makes and posts assignments for action. The last part of the value stream meeting is a review of overdue items still pending and assignments due today. This proceeds in the same way for the value stream as for the department, except all the value stream's support group representatives are regular attendees at the meeting and are potential recipients of assignments.

In an example of process trumping structure, the value stream leader acts as though the support group representatives report to the value stream. In some places, they do; in others, they do not. The leader makes assignments directly to the support group members and holds them directly responsible for getting their assignments completed on time.

If issues arise in these relationships, the value stream leader takes them up with his or her peers and perhaps with the plant manager (or equivalent superior). As organizations become better educated and aligned about Lean as an operating philosophy, the priority of functional projects rarely overrides immediate intervention on matters that threaten the expected flow of production or hold up measurable process improvements in daily operations.

Daily Accountability Exposes and Solves Problems Quickly

In the case involving drawers and locks, the Lean production process exposed problems that used to be covered up by rework occurring at essentially untimed workbenches. The Lean management system provided the tools and setting to get the problem resolved in a quick and accountable fashion.

Further Note on Task Assignments and Follow-Up

The supervisor at the department board and the value stream leader at the value stream board are free to make assignments from sources other than reasons for misses on the production tracking charts (as noted earlier in this chapter). Gemba walks with individual leaders, results of routine audits, and the leader's observations of the area can all result in task assignments. But assignments are not necessarily a one-way street. Subordinates or support members can bring requests for help or support to the meetings. In effect, they ask the supervisor or value stream leader to assign tasks to himself or herself where those tasks are best carried out by one in a position of more authority or with access to more or different resources.

In the third-tier meeting, when the value stream manager focuses on yesterday's visuals and asks for explanation, he or she sends a powerful message about focusing on the process. When he or she assigns tasks to respond to misses or opportunities to improve, a powerful message is sent about the importance of cycle-by-cycle performance and improvement. When red-dotted items are asked about in the meeting, it sends a powerful message about taking action, *now*. What happens, or fails to happen, in these meetings has widespread effects on what people in a value stream come to think of as important. Two contrasting cases illustrate this point (Case Study 5.2 and Case Study 5.3).

CASE STUDY 5.2:   RED DOTS INDICATE OVERDUE ASSIGNMENTS, BUT DON'T SHOW THE REASON

Consider these cases illustrating the rule of arbitrary green/red assignment. The Lean leader in a plant had an assignment on the accountability task board to deliver a Lean-related training session to a group in the plant at 2 p.m. on a Monday. At 1 p.m. that day, a fire broke out in the dust collection system for an aluminum grind-and-polish operation. Much of the building filled with white smoke. The plant and offices were evacuated for several hours, thoroughly disrupting the day's planned activities. Needless to say, the Lean leader was unable to deliver his training session. What was the result at the next day's task review meeting? He got a red dot!

In another case in the same facility, the operations VP began using a variation on the task assignment board to post and track projects assigned to his staff members. The assignment matrix was ruled off by weeks rather than days, reflecting the meeting's weekly schedule as well as the scope and scale of the projects. The tasks on the Post-its were the projects' agreed to weekly milestones.

The staff members liked the system because they could readily tell what all the projects were, and whether and when a project would have an impact on someone else's area. These were longer-term projects that usually extended over 60 to 90 days or more. Recognizing that, project owners were allowed to move milestone tasks from week to week for milestone dates more than a month from a given week's meeting. Milestone dates within four weeks of the meeting day were frozen so actual versus expected could be seen.

The staff member in a recently created position, responsible for sourcing materials and purchased parts, had several projects involving suppliers. These responsibilities had gone untended prior to filling the new position. Many of the sourcing staffer's project milestones collected red dots week after week as he pursued negotiations with the supply base on quality, cost, consolidation, lot size, and delivery frequency.

After using the new visual weekly project tracking system for several months, the VP and staff came to the conclusion they had underestimated the time it would take to conclude these projects. They agreed to "refresh" the board using the information gained from the actual versus expected red dots to reset expectations for how long this kind of work was likely to take. The sourcing person had been doing excellent work, the likes of which the company had never undertaken before. The operations staff as a whole was moving down the learning curve together in this area. There was never a hint of blame, shame, or recrimination arising from the red dots; instead, all involved treated it for what it was, a reflection of real-world events and valuable learning experience. Coincidentally, this experience of the senior operations staff was a powerful example for calming the anxieties of their subordinates, who were about to begin getting daily task assignments and green/red follow-up coding.

CASE STUDY 5.3: SUPPORT GROUPS CAN HELP SOLVE LONG-STANDING PROBLEMS

An outstanding example of support groups responding to the needs of the value stream occurred in a case in which a product line was added to a recently formed mixed-model progressive-build flow line. (This was the same cabinet assembly line where the operators in the drawer area brought in binoculars in Case Study 5.1.) The product line in question had been fabricated and assembled in a different plant since its introduction a dozen years earlier. In its original plant, individual operators working by themselves assembled units at workbenches, building the cabinet from start to finish. Now, the units were being built on a balanced, takt-paced assembly line. When more than one or two of a particular model of these units came down the line, things stalled badly. The production tracking chart showed missed pitches for the entire run of these products.

The cause of the flow interruptions was well known to those who had previously worked with the product. A particular style of drawer did not fit correctly and the cabinet's lock mechanism was balky. Each of these problems required extra work to make the units fit and function properly. When the units were bench-built, these models just took a little more time. The individual builders knew how to make the repairs. On a line paced at a 36-second takt time, these deviations caused missed takt cycles every time these problematic units came down the line.

At the value stream meeting the day after this problem emerged, the value stream leader assigned engineering and quality to deliver for the next day a plan to resolve the problems. They did so and were assigned to carry it out. Within two weeks, both problems of a dozen years' standing were permanently solved; the units now run smoothly down the line at takt time regardless of drawer style and with or without locks.

It's Not about the Boards!

It is not unusual to hear people say they want the visual elements in the Lean management system to have a consistent look, "so when people move from area to area or plant to plant, everything looks the same." In fact, the appearance of information boards and other examples of visual controls is far less important than how they are used. What's important is leaders' commitment to establishing standards and expectations, using tools to identify expected versus actual, and following through on what they find.

Every application of the principles that drive visual controls is an invention, for example:

1. Forms or charts
2. A units-produced counter
3. Bicycle flags identifying minimum and maximum levels in a queue
4. Audible alarms
5. Beads strung on a wire, as in billiards scorekeeping, to represent accumulating demand in a subassembly or feeder area
6. Tags on an inclined track or in pigeonholes

Insisting that every visual control should look like every other one misses the point. Yes, processes need to be systematic and standardized and the visuals need to fit these criteria as well. However, what is most important about visual controls is that leaders are committed to using them effectively. Where local process characteristics dictate varying from a visual standard, so be it.

Early on in the use of team information boards, one plant manager saw a board he liked (it had been carefully measured, rows and columns marked off with 1/8-inch automotive pinstripe tape, labels and date numbers engraved at a sign shop; the whole thing was impressive to see). He ordered copies for each of the value streams in his plant and decreed that all value stream leaders begin using them immediately. (It is worth noting that his background included significant responsibilities in national brand consumer goods production in multiple plants around the country. In this prior experience, everything was standardized across sites.)

Early Boards as Experiments

As it turned out, the design of the board's daily task accountability matrix was a winner, duplicated elsewhere. The rest of the board was very much a first draft, though cosmetically pleasing. Most of it has been revised. The copies that were ordered were soon festooned with paper charts, some of which were there at the direction of the plant manager after he realized the design he ordered did not reflect the data on which he wanted his value stream managers to focus.

Standardization is generally a good thing; standardizing before having had enough experience to work the bugs out of a design is not. If you are able to view as experiments the different approaches to even the same visual controls, you preserve an appropriate sense of learning by experience. In addition, allowing local experimentation gives leaders who have never before worked with visual controls the experience of having to think through what they want to measure and track, and why. That is valuable indeed.

As the experiments proceed, enough of a consensus will emerge about which design elements have proved the most useful to incorporate into a standard along with expectations for how the visuals should be used. That is the time to standardize. Keep in mind, however, that standards reflect only the current best practice. As further experience and insights accumulate, expect, indeed, encourage, the standards to change as a reflection of improvements.

Accountability Boards and Geographically Dispersed Locations

The most effective accountability boards are on permanent display, what you could call *visually persistent*. They are always there, in or close to the work area. Task assignments clearly show who is supposed to do what by when, and with green/red coding, whose tasks were done on time and whose were late. The meeting leader assigning tasks and the people getting the assignments stand face-to-face and can look each other in the eye. All of that adds up to the accountability in the accountability process.

But what do you do if the locations of the work and people involved with a tier three or tier four meeting are geographically dispersed? Units reporting into a tier three or four accountability meeting might be in different buildings, different cities, different states or provinces, different countries, or different continents. Some organizations rely on various forms of video conferencing to span these kinds of distances. If you have these resources, the people making and receiving task assignments can at least see each other. But the essence of the accountability process remains the accountability board: who is supposed to do what by when, and was the task completed (green) or not (red).

The accountability board in these circumstances will usually have to be electronic, often an Excel file reproducing the names-by-dates matrix of a physical board, with assignments appearing as name-by-date cell entries. As the meeting leader goes through assignments due on the day of the meeting, the cells can easily be coded green or red.

I have seen arrangements like this work well even though the accountability board was visible only during the meeting and only to those who dialed in to the conference call for the session and opened the accountability board Excel file. The effectiveness here rests on the meeting leader's discipline and his or her grasp of the accountability process.

Flat-Screen Monitors

I have seen several organizations that have chosen to put their Lean management visual tools on the internal IT network so they can be displayed in many places at once on flat-screen monitors. For some reason, this has seemed like a good idea, though one that I do not recommend.

First, the information has to be transcribed from a paper form, reworked for the network. But you might say, why not type the original directly into a database? Fine, if everyone's typing is good enough to capture what they wanted but didn't have, or had but didn't want. In many production positions people have not been hired with an eye toward their keyboard proficiency.

Information morphs from personal when you've written it down on a tracking chart, to impersonal, devoid of the fingerprint factor, when it appears on a network monitor, often after having been edited (or "cleaned up") by someone else for style or field width constraints. The screens seem often to be mounted high up where it is difficult to read, for example, the description of an interruption or delay. And, where the screens are programmed to show one chart at a time, they repeatedly cycle through many visuals, again making it difficult to absorb important detail.

In their favor you could say the flat-panel displays look good to visitors—unless the visitors include your Lean sensei.

When geographic dispersion requires it, use the network to connect people with visuals they would otherwise not see. But when people are in the same location, a more direct and effective approach is simply to use an accountability board. It may be low-tech, but you can count on it to be effective. Further, it is my recommendation to preserve the production tracking charts on which process problems have been captured in a handwritten note by someone who was there at the time. That is, save the paper charts rather than transcribing them into a computer file. If later someone needs to follow up on an observation, he or she can sort through the paper charts in far less time than it would take to transcribe every tracking chart for several months. Transcription in this case is waste, both working ahead and overprocessing.

The "Vacation Paradox" and Capacity for Improvement

Lean production applied in any value producing process, e.g., services, administrative processes, manufacturing, healthcare, is fundamentally an improvement system, but getting it to work that way can be difficult. Suppose you have a limited number of engineers and other specialists available for assigning to projects. You may wonder how all these continuous improvements will come about. As it turns out, the answer is in the latent capacity for improvement, developed in the daily task assignments in the second and third tiers of the three-tier meeting structure.

Supervisors who came up through just about any kind of conventional organization in any sector of the economy will usually tell you they have no time during the day to do anything but fight fires and expedite, plus sitting in meetings and doing other non-value-adding work that has landed in their laps from other departments. As the Lean conversion stabilizes, and especially as supervisors begin to follow their own standardized work, they find themselves spending less and less time fighting fires. In addition, their standardized work not only makes it clear what they are supposed to do, but by implication it also makes it clear what they are not supposed to do. That is, they are not supposed to be running here and there doing the jobs of quality, material supply, and engineering. Instead, they are supposed to use those cell phones and two-way radios to notify their support groups when things go wrong outside their production area. And, in the process of creating clearly defined standardized work, much of the extraneous work from corporate and other groups should have been greatly reduced or eliminated entirely (Case Study 3.1 in Chapter 3).

Improve-the-Business Capacity

Overall, this has the effect of freeing chunks of time in their daily routine. Standard work only accounts for about half of a supervisor's time, often less in administrative and technical-professional processes, and even team leaders should have about 20 percent of their time available and unassigned in their standard work. The intent is to put at least some of this discretionary time to use on activities to improve the business. The source of these activities comes from the daily task assignments made at the tier two and three daily accountability meetings. All of this provides the vehicle for transforming the 100 percent run-the-business focus to a shared one that includes regular work on improve-the-business activity (Case Study 5.4 and Case Study 5.5).

These small, one-step-at-a-time daily assignments give frontline leaders the experience of doing more during the day than just running the day-to-day business. Many frontline leaders have had little or no experience with project management. The daily task assignments represent project management work breakdown structures on a small scale. These assignments show how small steps can lead to significant outcomes when disciplined accountability for execution is part of the picture.

Besides that, leaders learn that these steps can easily fit into a daily schedule, especially when an area is operating in a stable state. Further, these cumulative improvements frequently do not require teams of technical specialists. I have a name for this phenomenon of suddenly having the time to do more, to work on improvement steps in addition to run-the-business activities. I call it the *vacation paradox.*

It works like this: when I am about to go on vacation, especially if it is a fishing trip, which means leaving my family behind, I want to be sure there are no loose ends at home—or at work—during the few days or week

CASE STUDY 5.4: IGNORING INFORMATION ON BOARDS VERSUS USING IT TO IMPROVE

In the first case, team information boards had been established for each team leader's area as part of an important new product's showcase Lean production area. The boards were neat and uniform, and lots of emphasis was given to keeping the entries on them current; the area had lots of visitors from VIPs and customer groups. So, the team leaders all learned the importance of maintaining their boards and did a good job of keeping the entries current, including the entries on their production tracking charts.

As with many new product introductions, the design for the production process had looked good on paper. Once it was installed and operating, it became apparent that several aspects of the process should be changed. Project teams were formed to make the changes, and the area's performance improved. In fact, virtually all of the changes made in the area for its first 18 months of operation came from project teams working on fairly substantial changes to the layout.

Near the end of this time, I happened to ask a team leader what he thought of the information board he so dutifully maintained. He snorted and said derisively that the board was a joke. He had been recording reasons for misses for nearly two years, he said, and not a single thing had happened as a result. He still faced the same problems day in and day out regardless of how often he wrote up problems in the reasons for misses section of his production tracking charts. "I'm going to quit filling the darn thing out," he said. And he did. Nobody said a word to him about it until a new value stream manager and supervisor were named for the area. Until that time, the main focus had been "hitting the numbers" for schedule completion and cost. Although this plant has performed satisfactorily, there is a sense that it could be doing better than it has.

In a contrasting case, I was touring an assembly area in a plant that had undergone and continued to undergo changes designed to make its layout and processes leaner and leaner. The plant Lean team had introduced production tracking charts and the daily accountability process recently, as part of a Lean management implementation.

I stopped and asked an operator (who had no idea who I was) if she had a production tracking chart. She said she did and showed it to me.

It was filled out, noting where production had missed its hourly goal and a reason why.

I asked her what she thought of the chart. Her face lit up as she proceeded to tell me what a good thing it was. "We've written down things that have frustrated us forever," she said, "and they've been taken care of, just like that! It's a lot better around here now with these forms," she said. When I looked at the department's information board, I could see the list of top three flow interrupters and actions underway to resolve them. Clearly this was an area that paid attention to the health and improvement of its processes. This plant has been outperforming its goals at the same time it has been absorbing substantial change.

CASE STUDY 5.5:   THE POWER OF MANAGING ASSIGNMENTS VISUALLY

I had the opportunity to work with plant maintenance groups in two geographically separated companies in different industries, both of which were integrated mining and chemical operations. In both companies the production process occurred in three stages. Mining the ore was the first stage. Preparing the ore for processing came next. Third was a continuous process chemical operation that produced the end products.

In this kind of process, heavy maintenance activities in either ore preparation or the chemical process can only occur when production is completely shut down. In both companies this was referred to as a turnaround. In both cases here, the turnarounds were in ore preparation and involved tearing down, upgrading, and rebuilding major assets in the ore preparation stage of the production process. If a maintenance turnaround misses the scheduled due date, production remains shut down until maintenance puts the assets back in operating condition. It should be no surprise that executive, operating, and maintenance management put a great deal of emphasis on completing the turnaround work on time and done right. Even so, in both cases, turnarounds had never met their schedules because of finding the unexpected as the maintenance crews got into the inner workings of the processing assets.

Both organizations had been exposed to the tools and practices of the Lean management system, but I was surprised by what I saw.

They did not have much in the way of process-focused visual controls, nor was there well-defined standardized work for either the frontline workers or leaders. But, both maintenance groups were using accountability boards to follow up on tasks in the turnaround plan that were at risk of not being completed on time, and for making assignments for tasks that emerged from the maintenance teardowns. The names on the accountability board included all the maintenance supervisors and their managers, all the relevant support group supervisors and their managers, the operations supervisors and their managers, the plant manager (who attended many of the meetings), and the site manager for the whole end-to-end operation (who attended occasionally). This group of line and support group leaders (a maximum of 12 when all attended) met in front of the boards once a day in one organization and twice a day in the other to make assignments and code previously assigned tasks green or red depending on whether or not they had been completed by the due date.

The leaders of both maintenance groups said they were convinced that having these critical tasks assigned to specific people, posted in public, with publicly visible green and red coding, reviewed at least daily by peers and superiors, made all the difference in getting their turnarounds completed on time, for the first time, in both companies.

I am gone. So, I scan my world for things that need to be tied up before I leave. The last thing I want is to return home and face an unhappy family because I forgot to look after a detail that ended up inconveniencing them! The same is true at work. I want to return to no more than the usual accumulation of email and voice messages. I certainly do not want to return to find I left a smoldering situation or untended assignment that burst into a full-fledged problem while I was out on holiday.

In short, during these prevacation times I get lots more done at home and at work than I usually do. I do not transform into a whirling dervish of activity, but I do use my time more efficiently. That is the vacation paradox: there is a lot more time to get small things done when they are clearly identified and scoped and there is a reason to get to them. Avoiding unhappiness in my family or boss is one reason to get to them before vacation. Another is having tasks done and coded with green dots rather than red for the peace of mind it creates.

Impact of a Daily Few Minutes

Imagine the impact of the team leaders and supervisors in a value stream devoting as little as 15 to 30 minutes every day to improvement tasks! The capacity to do this work has been present all along. The unstable, firefighting nature of leaders' roles in a conventional production environment, plus all the other work outsourced to them from internal departments and groups, had demanded all of their remaining time and energy for meeting the daily production schedule. In an increasingly Lean environment, as production stabilizes, it is possible to tap this latent capacity, though it takes time before the extra work ceases to seem like an imposition in an already busy day. When that happens, it is possible to say that an organization has developed the habit of daily improvement.

The power of this process can be surprising. Leaders and front-line workers around the world have commented on the effectiveness of the daily improvement assignments keyed to recurring causes of interruption. Problems that had seemed as much a part of the workday as coming through the door in the morning had been worked into oblivion step-by-step and were gone. Operators, as in the job-by-job area noted in Chapter 4, were happy to have a vehicle for getting their problems solved. And leaders were initially amazed at the results the visual, accountable, daily focus on correction and improvement produced, as in the drawer fit and lock function case study (Case Study 5.3) in this chapter.

Accountability in Office Processes

In repetitive manufacturing, it makes sense to focus on accountability daily. Variations can come at you fast in a high-volume factory. That is also true in high-volume office processes such as order management/customer service. But where office processes do not cycle so quickly, for example, in executive and staff groups, IT development, new product design and engineering, aspects of legal, HR, and compliance activities, weekly accountability sessions can be a better fit. The objective is to produce improvement, not to hold meetings for their own sake.

Summary: Daily Accountability Improves Processes

Daily accountability is a vehicle for ensuring that focus on process leads to action to improve it. The structure of the daily accountability process is straightforward, a series of two, three, or four brief meetings to review what happened yesterday (or in the last week) and assign actions for improvement. These are time-paced, stand-up meetings in or near the work area floor that emphasize quickly resolving or investigating to the next level interruptions in the defined production process. Standardized, time-paced, green/red-coded, weekly plant or business unit accountability meetings follow up on task assignments due this week related to larger-scale problems, or review completion status of milestones for the week in larger-scale improvement projects. The purpose of these meetings is not communication of information someone somewhere in the organization thought others ought to hear, but rather are we getting done what we set out to accomplish, and if not, what is the recovery plan and expected completion date?

Daily accountability is the vehicle for interpreting the observations recorded on the visual controls, converting them into assignments for action, and following up to see to it that assignments are completed. As with the other principal elements of Lean management, daily accountability relies on disciplined adherence to its processes on the part of those who lead the three-tier meetings. When this discipline is present, leaders follow their standard work. Leading the three-tier meetings effectively is an important part of leaders' standard work, which is to make it clear that maintaining visual controls is vital, expected, will be closely reviewed, and leads to action.

The daily accountability process also provides a vehicle for introducing and modeling the basics of project management for those leaders who came up through the ranks and may not have had exposure to these tools. And, the process definitely reinforces the connection of support groups with the manufacturing, service delivery, or other frontline process by involving them in daily accountability assignments to resolve issues that interrupt or delay the value-adding frontline work process, and quickly. This is true whether or not support group members have a formal reporting relationship to the value stream.

By reinforcing process focus and driving improvement, daily accountability actually creates increased capacity for improvement via the vacation paradox. Finally, the green dot/red dot convention brings visual accountability for expected versus actual into concrete reality for a value stream's leaders and staff members. The effect is a powerful reinforcement of this theme running through Lean management. See also the Lean management standards and gemba worksheet for the daily accountability process in Appendices A, B and C.

Study Questions

1. Have you had experience with charts or lists of problems that did not result in actions for improvement? If so, why do you think the lists were ineffective?
2. Is there a sense here that when someone gets assigned to an improvement task, it may never be acted on? If so, why do you think it happens?
3. How easy is it to tell what is being done by whom to address problems that interrupt or delay daily work at the front line? Can you tell when an answer, next step, or solution is due? Are the due dates taken seriously?
4. What gets the most emphasis here: hitting the numbers or improving the process?
5. Is the investment in start-up meetings here, if they exist, worth the time they take in terms of improving performance? Do people seem to be paying attention to the content being covered? What might improve them?
6. How much of a sense is there here of attention being paid through several levels of leadership to problems that disrupt daily work at the front line?

Chapter 6

Lean in Administrative, Technical, and Professional Work

Lean is no longer just for manufacturing. The roots of Lean are deep in the shop floor, of course, beginning with Henry Ford. But, in recent years, Lean thinkers have found the principles and many of the tools of Lean are just as effective in administrative, technical, professional, and healthcare processes. (For the purposes of this chapter, the terms *enterprise Lean* and *enterprise business processes* encompass Lean beyond manufacturing.) And, Lean management approaches are needed to sustain and extend Lean in enterprise business processes just as in manufacturing.

This chapter illustrates some Lean and Lean management applications in the world beyond the manufacturing floor. It describes relationships among process problems in the full range of enterprise business processes, solutions by application of Lean tools, and sustaining the gains with Lean management applications. A second subject is resistance to change. Resistance is a potential issue in any Lean implementation, although its background is often different with Lean office initiatives. Accountability can spark resistance for office workers; usually, overcoming resistance simply takes consistent execution of Lean management.

The chapter's concluding subject addresses the organizational and political environment in which enterprise business processes operate.

Enterprise Lean faces particular pitfalls characteristic of its environment, specifically regarding leadership alignment. Lack of alignment can be a problem in any Lean implementation; in enterprise Lean, it can be particularly troublesome. This final subject details a proven process improvement project design built around value stream mapping of end-to-end, cross-functional business processes. The design is a repeatable, structured, high-accountability, high-alignment process. It involves both top-down and bottom-up perspectives to identify and eliminate waste from these often complex, contentious, and tangled processes. The outcomes are consistently better, faster, and more efficient delivery of value to the ultimate customer. The design anticipates and neutralizes problems with organizational alignment and builds leaders' experience and engagement with the enterprise Lean initiative.

Lean Management in Enterprise Business Processes

Most things of value, whether a service, a drawing, new knowledge, or a manufactured good, are produced as the result of a process. As a process improvement system, Lean applies to any process. For this discussion, consider linear or step-by-step processes. Some examples are handling a customer service request, establishing a new customer account, replying to a request for proposal, updating catalogs, making changes to terms and conditions, scheduling patient appointments for ancillary services, modifying a standard offering per a customer's request, manufacturing a product, or closing the books at month's end. Many problems familiar to Lean practitioners crop up in these kinds of processes, and Lean principles apply in eliminating them. Table 6.1 shows just a few examples of problems in enterprise business processes and the application of Lean tools to improve flow and reduce waste.

Just as with manufacturing, implementing Lean tools in enterprise business processes is the easy part; the tools require no more than 20 percent of the effort, often less. Sustaining the Lean implementation takes most of the work. As in manufacturing, this work typically remains undone, even in the best examples of enterprise Lean tool implementations. For that reason, sustaining enterprise Lean implementations requires the parallel implementation of a Lean management system. Table 6.2 shows Lean management applications that could be used to sustain the Lean responses to business process problems.

Table 6.1   Office Process Problems and Lean Solutions

Process Problem	*Lean Response*
1. Large batches produced between steps in administrative processes increase wait time, slow throughput, and conceal defects.	• Create defined flow lanes to control work in process (WIP) between processes. • Or, link process steps in a cell. • Define and implement separate flow paths.
2. Uncontrolled release of work into an engineering process causes imbalance, bottlenecks, and backups when separate flow paths are not defined for jobs with substantial variation in work content.	• Define separate flow paths and implement logic for controlled release of work into each path based on calculated or estimated work content. • Track actual versus expected work content to improve accuracy of estimates and calculations.
3. Misunderstood requirements between process steps result in rework in the receiving step to correct defects or seek missing information from the supplying step.	• Link processes where possible. • If not, kaizen to design the handoffs at boundaries, and define mutual expectations with handoff-ready checklists.
4. Downstream processes start early in an attempt to speed up throughput only to rework their "early" efforts when final input to their process differs from what they anticipated.	• Kaizen to create release-ready checklist to be completed before work is released to the downstream process.
5. Unplanned downtime in the IT network causes lost production, missed delivery dates, and overtime.	• Create user downtime log. • Regularly review with IT. • Visual status of tickets. • Visual status of system improvement work.
6. Wide variation in method and sequence among people in the same jobs doing similar tasks results in variable quality and productivity.	• Define standard sequence of steps. • Document standardized methods. • Document expected productivity, track expected versus actual to capture breakdowns for accountability task assignments.
7. Many people order supplies, resulting in overstock of some items, stock-outs of others, and mystery items nobody seems to use.	• 5S supply areas. • Put supplies on kanban replenishment system. • Define reorder process and responsibility.

Table 6.2 Lean Management Applications for Process Problems

Process Problem	Lean Response	Lean Management Applications
1. Large batches	• Flow lanes or flow paths or a cell— virtual or physical	• Define processes and flow paths • Leader standard work to monitor adherence • Production tracking chart to identify problems • Accountability processes to understand and address causes
2. Release of work not controlled	• Release work to flow paths based on estimated work content and signal for "next" • Define processes and flow paths	• Defined process for controlled release of work • Leader standard work to monitor timing of work release • Production tracking chart to identify variations from expected work content, delays, interruptions
3. Upstream, downstream needs not understood	• Link processes, define needs, provide feedback • Create and use release-ready checklists prior to release of each unit of work	• Joint start-up meeting of both steps • Leader standard work to monitor timing and quality of work to be released and related problems at receiving step • Tracking chart at receiving process to record agreed timing, completeness, accuracy of work received • Accountability processes to understand and address causes of supplier and customer steps
4. Starting too early	• Release-ready checklists	• Leader standard work to monitor use and effectiveness of release-ready checklist • Defined process to control release of next unit of work based on estimated time to complete the unit

Table 6.2 (*Continued*) Lean Management Applications for Process Problems

Process Problem	Lean Response	Lean Management Applications
		• Production tracking charts to record rework prior to release, quality and timing problems at receiving process
5. Unplanned downtime	• Visual downtime log with reasons for delays	• Leader standard work to monitor use of log • Visuals with pending and current tickets and expected due dates for jobs • Joint reviews with IT to rank problems and maintain accountability for timely resolution
6. Variation in method and sequence	• Standard sequence, methods, expectations	• Standardized work or procedures • Visual controls for progression of work through the standard sequence • Visual controls within steps to capture problems • Visual controls at handoffs for progression of work through standard sequence and to capture delays, defects • Leader standard work to monitor adherence to standardized procedures • Accountability processes to understand and address causes of variations and problems
7. Overstocks, stock-outs, mysteries	• 5S; replenishment order points (kanban or pull system) for supplies, materials	• Defined responsibility for pull system audits with posted results, accountability assignments as warranted • Accountability process to understand and address causes for identified misses

Lean: Not Just for Manufacturing Anymore

Enterprise Lean first spread to administrative functions from the experiences and lessons Lean manufacturers learned in their factories. Now Lean initiatives can be found in diverse sectors of the economy beyond manufacturing, in organizations whose products are solely knowledge or services. Among many examples are finance, insurance, government, healthcare, software, pharmaceutical development, sales, order fulfillment and distribution, customer service, and education. An increasing number of organizations have moved beyond asking, "Will Lean work for me even if I don't manufacture anything?" Instead, the question now is, "How do I apply Lean to improve my business processes?"

For the practitioner, Lean and Lean management apply nearly transparently in processes like these, just as in manufacturing. But unlike in manufacturing, disciplined process analysis and intentional design are rarely—if ever—applied to enterprise business processes. Instead, many office processes have been left to evolve over time, usually with little examination. I have never been in a factory that did not have engineers concerned with process design, even if not always a Lean design. I have rarely been in an office in which engineers or anybody else was involved in designing or refining business processes.

For these reasons, and from experience with nearly 200 Lean enterprise projects, I am convinced there is as much to be gained in improved throughput, quality, productivity, and engagement in offices as in factories. Initial gains from Lean enterprise projects commonly reach or exceed 50 percent reductions in total cycle time, process time, and errors and rework, as well as revealing further opportunity for improvement of the same magnitude, just as in manufacturing.

Resistance: Accountability and Visual Controls

Visual controls are important in any Lean management implementation. In enterprise business processes, the progression of work—as from one department or specialist to the next—typically cannot be seen because handoffs occur via work flows or emails. For this reason, visuals might be even *more* important than in the factory, where flow and interruption are usually in plain view. In nonmanufacturing business processes, work typically progresses through the IT network, hidden from easy observation (except in healthcare waiting rooms and emergency department hallways). You can see people at computers or on the phone, but you cannot see the work.

It is typical in enterprise Lean projects to find that half, 50 percent, of work time is spent on work-related but non-value-adding activities. These are things like doing rework to correct errors or changes in requirements or inputs, making requests for missing information, seeking clarifications, and waiting for approvals or replies.

The amount of rework and non-value-adding activity in office processes may contribute to the tendency for office workers to resist the increased accountability that comes with Lean and Lean management. Compared with manufacturing, applying time-based expectations for performance (and measuring actual versus expected output) is not as obviously a fit in enterprise settings. Business process work often involves considerable variation in the nature and duration of work content from one unit of work to the next. And, workers in these settings are simply not as used to being held accountable to the same extent as production workers in manufacturing. Many enterprise process tasks are treated as specialized, only to be performed by "specialists." With office work framed this way, expected rates of output and other measures of yield and productivity have not been widely applied, to individuals or to the team.

Making Performance Visible—and Comparable

Lean thinking changes many of these assumptions and practices. Visual controls in offices, as in factories, raise the level of accountability for everyone involved in a process. So do expectations for productivity. Cross-training and cellular designs shrink the number of specialists and bring common standards to many areas of performance. Of course, this allows ready comparison of performance among people and teams, another form of accountability. Any or all of these factors can contribute to Lean initiatives facing greater initial resistance in office areas than in factories. What should you do when faced with initial resistance?

Stick with it! Demonstrate by your actions that people are taken seriously when they record problems on the visuals or suggest process changes. Act on the process misses and interrupters recorded on the visual production tracking charts. Work quickly to eliminate real problems and implement constructive suggestions. Explain why some problems or suggestions will not be taken up. (These often represent requests to return to the previous mode of operating. As such, they represent teachable moments, opportunities to review a point or two in the rationale for Lean.) As you respond to and eliminate problems people have brought to your attention, you will also be eliminating some long-standing sources of frustration for those who do the work. Some people will not be happy about doing things differently. But, when you follow through on problems and suggestions, people will experience the benefits, and the resistance of most of them will fade. Similarly, green/red coding of visually posted assignments and project milestone tasks shows progress being made, and credits people and teams for getting things done.

These actions are powerfully reinforcing. They answer the frequently unasked, but widespread and crucially important question: "What's in it for me?" You can still have conversations about Lean's benefits to the business. But you can—and should—also have conversations comparing the old way, shot through with waste, waiting, rework, and bottlenecks, with the newly improved approach. Workers experience less frustration and more success in an environment with improved flow, fewer interruptions, and information that is timely, accurate, and complete. These all add up to having a better day at work. And, having a better day at work is at least as (and usually far more) important to people on the front line as the benefits to the business. Resistance usually dissolves among all but an unhappy fringe (up to 10 percent) of people. You or they may well decide they'd be better off working elsewhere. These experiences are reflected in Case Study 6.1, involving marketing, engineering, and manufacturing database departments, and Case Study 6.2 in a billing services group.

A simple but important step in reducing non-value-adding effort in office processes is to make it visible. Value stream mapping the process is often a good first step to show the waste. But it is not unusual to find disagreement on what to map, and on how the process works among those performing the same tasks in the process! In these cases, it is necessary to begin by developing agreement on the steps and sequence that constitute the process. Once agreement is reached, it is possible to map and then visually track the progression of work, and to see opportunities for improvement. For an example of a visual used to track progression of work through an engineer-to-order/database updating process, see Figure 6.1.

When work moves between steps in an office value stream through the IT network, visuals allow you to ask and answer questions like these:

■ Where is a given unit of work in the sequence of steps?
■ Has each step been completed on time and in the agreed sequence?
■ If expected throughput time in a step has been exceeded or missed, what was the reason for the miss?
■ If the process is interrupted at a given step, what is the recovery plan?

CASE STUDY 6.1:   MY PROCESS,
YOUR PROCESS, OUR PROCESS

A product data group created the marketing and manufacturing databases and "smart" product graphic symbols used by customers to create layouts and then order products. Together, this database work was the longest item on the new product development critical path. Standard time from start to release was nine months, often delaying new product commercialization. The process had for years resisted repeated attempts to improve it. As a last resort, the area manager asked for help from the office Lean team.

In doing background work, the Lean team found marked differences in how the database people did their work. Some seemed to start in the middle, some at or near the beginning, and no two people seemed to follow the same sequence of steps. In a series of intense workshop sessions with the Lean team's help, the database people agreed on a defined set of steps. In following sessions, they hammered out a sequence in which they would work through the steps.

One more workshop session produced on a large whiteboard a matrix of projects (as the rows) by process segments, and steps within segments (as the columns). With steady follow-up, the group came to follow the steps uniformly. Now it was possible for everyone to see the progression of work on each project, and daily stand-up meetings at the board enabled the team to mobilize help and develop an understanding of problems when a project got hung up at a particular step.

A note on resistance: "Hammering out" is not too strong a description of what it took for these technical specialist office workers to agree, however grudgingly, to try the new process. Two factors were at work that led to overcoming their resistance to change, one internal to the department, one from outside.

The outside factor was the failure of all previous efforts over many years to reduce the bottleneck represented by the database process and the costly delays it added to product introductions. Even after completing implementation of a new enterprise resource planning (ERP) software system that encompassed engineering, marketing, manufacturing, order management, as well as database, there was no improvement in database throughput. Database was now a corporate-level problem, with the attendant pressure on management to find a better way.

There was no change in the value of treating people with dignity and respect. Now, however, it became clear that finding solutions to the database bottleneck had to go forward even if people did not like it—and that did not constitute disrespect for the database group. So, the department manager pushed for change more insistently and with greater firmness than his predecessors.

The internal factor was the power of value stream mapping. The mapping sessions revealed a veritable spaghetti bowl of rework, over-processing, and disconnected schedules and priorities, starting some work too early and other work too late; the map was nightmarish. It is said that reasonable people, given the same information, will reach generally the same conclusions. That ultimately was the case here.

Supported by the department manager, the Lean team facilitators firmly but respectfully forced the database group to face the reality of their work processes. It was clearly reflected on the current state value stream map the database group had created, supported by the Lean team members. As part of the workshop, the database people had gone through Lean training. Now, in the value stream mapping sessions, the Lean team facilitators reinforced and interpreted the Lean lessons. Between their manager's insistence and the Lean facilitators' support, enough of the database people realized they simply could not continue asserting their process could not be improved. They agreed, not without reluctance, on a future state that proved to be the key. In a matter of two years, the new process led to quadrupled productivity in database, and over two-thirds reduction in cycle time. Now, this way of operating is taken as a given—and a point of pride—in the group.

Three years later, a similar process was implemented in the group responsible for modifying the databases when standard products were customized, or engineered to order, at a customer's request. The database group consistently experienced rework because of incomplete or inaccurate inputs from the engineer-to-order team. And, the engineering group was chronically late getting the new specs to the database team, which then had to rush, in addition to performing the frequently required rework. Often, the custom products missed the promised shipment date. Customers were unhappy, and so was everyone working in the process!

This time, Lean thinking had spread widely enough that there was little resistance from the engineers and database people to the idea that their process could be improved.

In a Lean project, the two groups worked together to create a value stream map of the steps in their processes. They immediately realized their steps needed to be interspersed with each other, and that working in batches in department silos was the cause of much of the rework in the database group. After creating a future state process design and value stream map, the project team created a daily update board showing the progression of work through the engineer-to-order and database process, and established agreed dwell times through each step in the process (Figure 6.1).

In a matter of a month, late releases and deliveries stopped, as though by magic. Both groups now swear by the power of this method of visually controlling their joint process.

CASE STUDY 6.2: INTERRUPTIONS MANAGED, CUSTOMERS SATISFIED

A manufacturer's billing services department handled disputes arising from complex contracts involving the manufacturer, distributors, and customers. Terms and conditions could differ with different versions of contracts, for different years, various locations, and different phases of projects. Untangling distributors' claims was often a lengthy, detailed, and complex matter.

Billing representatives were in the middle between sales administration, which was responsible for the contracts, and the distributors. The four billing reps had been in their jobs for several years. They found their work challenging and, when they succeeded in resolving the disputes, satisfying. The reps regularly experienced frustration from being interrupted in the midst of researching and working through these complex business problems. The primary source of the interruptions? Phone calls from customers, distributors, and the sales administration department.

The phone system was set up so that incoming calls would trunk to a free line among the four reps and their department manager.

When all lines were busy, calls went to a department voice mailbox. Voice messages often languished for a day or more before someone got around to listening to them, leaving a message for the rep on the case. Reps typically waited to return calls until they resolved the element they were working on. On any open dispute, sales administrators and distributors might, and often did, complain to the director about the unresponsiveness of the billing services group. The director interrupted the manager to get on the complaint. She then interrupted the work of the rep involved. Discussions with IT about doing something different with the phone system had gone nowhere. The director felt she was at a dead end.

The director had been exposed to Lean manufacturing in an MBA class she was taking. She was interested, and got in touch with the internal enterprise Lean team. A Lean team member met with her and asked about the director's work problems. The director mentioned the area's frustrations with the endless interruptions, and their reputation for unresponsiveness.

The Leanster walked the director through the logic of timed release of work, where specific tasks are scheduled to be performed at specific times to maintain flow and meet planned capacity. When she asked the director how that might apply to her problem, a light went on for the director. The following week she calculated the daily average number of incoming calls, and the duration of a typical reply. That allowed her to calculate total time to return a typical day's calls. Dividing total time by the four reps plus the manager came to about 30 minutes per person. She figured if all the incoming calls trunked to the voice mailbox with a message promising a callback within an hour, it could work with each rep and the manager returning calls for 15 minutes twice a day. One of the five would handle all the voice messages on the hour, rotating every hour among the staff once or twice during their eight-hour day.

The director discussed this approach with the manager and reps. They agreed to try it for a week, and found that it worked. Callers were pleasantly surprised with the new responsiveness. Reps now rarely dealt with callers who were angry and frustrated about long response times. The reps could maintain the concentrated focus they needed to sort through the disputes.

The Leanster then suggested using a visual control for number of dispute cases in queue, average time to resolve them, and an accountability board for assignments to eliminate or minimize problems that delayed resolution. A brief daily stand-up meeting reviewing the visuals and assignments sustained the gains from paced release of work, and stimulated further improvement in the area.

Engineering to Order –Mfg and Mkt DB Project Board 9am Daily Standup

Proj #, owner # lines	Order entered	Eng edit	Need to clarify?	Eng design	Mfg database	Mkting database	Effectively applied	Release to mfg	Reason for early, overdue
Dwell time	Date	1 day		1 day	½ day	½ day	½ day	1 day	
T6F31 Rez 2	1/15	1/15 1/15		1/15 1/16	1/16 1/17	1/17 1/18	1/18 1/19	1/22 1/22	Pricing error found for expedite delivery
T6F68 Pam 3	1/19	1/22 1/22		1/22 1/23	1/23 1/23	1/24 1/24	1/24 1/24	1/24 1/24	Close to previous design
T6L88 Tom 1	1/19	1/22 1/23		1/23 1/24	1/24 1/24	1/24 1/26	1/26 1/29	1/30 1/30	Expedite price DB rebuild now fixed
T6A96 Phan 1	1/23	1/23 1/24		1/24 1/25	1/25 1/26	1/26 1/26	1/27 1/27	1/28 1/27	
Legend		In / Out	Late						Reason overdue: complete when occurs

Figure 6.1 Visual control for progression of work through an office process.

Just as in manufacturing, visuals in enterprise business processes allow you to identify and respond to variations in process performance. Misses in enterprise processes can become the basis of a task on an accountability board, are often documented on a Pareto chart at the end of a week, are frequently the subject of root cause analysis, and might appear as the problem statement in an A3 project to implement a root cause solution. Leader standard work sustains the building block elements of the Lean management system by focusing on executing the visuals, the accountability processes, and the other elements of Lean management. But the units of work in office processes can vary from unit to unit (for example, difficult versus routine requests for post-sales service, or first-time designs versus minor modification). Because business process work often lacks the repetitive tasks of some manufacturing settings, the percentage of a leader's time occupied by his or her standardized work may be less than that of his or her counterparts in the factory.

Enterprise Value Streams and Their Political Environment

Successfully implementing Lean in office and technical-professional value streams can be more challenging than in the factory. This is not because of greater technical complexity in office processes; Table 6.1 shows how Lean tools apply similarly whether in office or factory. (Lean management standards for enterprise business processes appear in Appendices B and C, along with the standards for Lean management in manufacturing.) The challenges for Lean success in enterprise value streams come from the more complex organizational and political environment encountered in cross-functional office processes.

Think of it this way. In manufacturing, there is almost always one person with responsibility for the plant and its people, processes, and performance. While many functions besides manufacturing are present in factories— safety, production control, supply chain, engineering, maintenance, human resources, quality, finance—everyone understands the plant's goals: ship quality products on time, at the right cost, and do it safely. Some in the plant might report elsewhere in the organization. But when plant performance is at stake, everyone is accountable to the plant manager, whether it looks that way on the organizational chart or not.

There is no parallel position in enterprise business processes. Enterprise value streams cross many organizational boundaries, cutting across the vertical "silos" of the functional organizational structure. Products or services get to customers from the efforts of sales, marketing, order management, product development, IT, finance, distributors and a distributor relations group, perhaps engineering and production, and often others. Some of these functions might report to a single higher-level executive, but even so, can you identify a single position with ownership responsibility for the end-to-end process, similar to the responsibility of a plant manager? Almost certainly not. For example, see Case Study 6.3.

Business processes almost never appear on the organization chart. Enterprise value streams, another way to refer to business processes, cross many internal boundaries. They rarely have identifiable, accountable ownership or dedicated resources and budgets. Cross-functional business processes typically lack their own measures. Indeed, improving an end-to-end

CASE STUDY 6.3: FIRST SEEING, THEN IMPROVING A BUSINESS PROCESS

In the value stream improvement project described in Case Study 1.3 in Chapter 1, the hospital's VP for ancillary services sponsored a value stream improvement project he thought would support his plan to expand a centralized scheduling pilot to cover all patient referral appointments from the hospital's outpatient clinics and physician practices (such as family medicine, oncology, women's health, and orthopedics).

The current state value stream map created in the project workshop by the frontline administrative workers from the practices and clinics led to a different conclusion. The map showed the centralized scheduling pilot was only effective on paper. In practice, it was shot through with waste: phone tag, delays, rescheduling, wrong instructions, and overall a source of frustration and dissatisfaction for patients. The centralized service did provide convenient single-source financial reports for the VP. But it appeared that frustrated patients increasingly were choosing to receive the prescribed tests and procedures at competing service providers in the area. The operational goal of the project sponsor was to schedule referral appointments during the patient checkout process, before patients left the clinic. His method, the centralized scheduling call center, was not achieving the goal and alienated the clinics and practices, and their patients.

The VP was right about the benefits. Scheduling at checkout would eliminate the multiple phone calls from central scheduling (and the resulting phone tag), rescheduling, and wrong or incomplete instructions to patients (e.g., fast for 12 hours before a lab test) that meant canceled appointments and caused the scheduling process to start all over again. If these frustrations could be eliminated, patients were far more likely to use the hospital's own facilities, allowing the hospital to better serve its patients and capture the ancillary service revenue it so badly needed. But he was wrong about centralized scheduling providing those benefits.

In the course of the workshop, the team members, who represented the hospital's several satellite clinics and its five hospital-based services (lab, imaging, cardiac rehabilitation, etc.), went into detail with each other about how they sent and received referrals. The hospital had recently installed a new IT system across all its units, including the ancillary services, clinics, and physician practices. One of the workshop team members, the main referral scheduling clerk in one of the practices, had discovered the system's new instant message feature and figured out on her own how to use it with a few of the services to request the referral appointment the patient wanted, and to confirm it before the patient checked out.

Everyone in the workshop had been trained on the new IT system, but the trainers, unfamiliar with the users' work processes, had not covered the instant message feature. Description of the instant message feature lit up the workshop participants. They designed a future state value stream map and process based on the newly discovered ability to instant message each other. The future state implementation included the clerical staffers themselves designing a training program for requesting and scheduling referrals, administering the training to their peers across the system, and keeping the training current with new developments. At last report, the new process was an ongoing success.

This project exposed a broken process, the reality of which had not previously been grasped. It enabled those who performed the work to examine the variations among them, and to recognize and capitalize on a previously unknown best practice. And it converted a neglected set of piecemeal practices that drove patients away into a smoothly and nearly autonomously functioning process, transforming a weakness into a competitive advantage. It is a great example of the power of enabling those who do the work to use Lean concepts and tools to examine what they do and how they do it. Coupling Lean concepts with the perspective of an end-to-end business process allowed these frontline staffers to dramatically improve the business process, deliver improved value to their customers, and improve efficiency in their own organizations.

business process often negatively affects the measures of individual departments, which are tuned for optimizing performance on functional measures. For all these reasons, cross-functional improvement projects often come up short. Projects run into internal boundaries. Ultimately, competing priorities and department or functional boundaries defeat the effort to improve the end-to-end process. Even if the end customers and the organization as a whole would benefit, internal boundaries and competing priorities stymie progress. If a given department's priorities are elsewhere, the value stream project is out of luck (Case Study 6.4).

CASE STUDY 6.4: DEAD ENDS IN CONVENTIONAL PROCESS IMPROVEMENT

My own experience provides several examples of the futility of conventional process improvement approaches. Long before I heard of Lean, I worked as an internal organizational development (OD) consultant in a Fortune 500 company. In my early years there, I was involved as a facilitator in many conventional large-scale projects aimed at resolving problematic performance in critical aspects of the business, such as the new product development process, accurately capturing and fulfilling customer requirements, the timeliness of the engineer-to-order special products process, or untangling difficulties with specifying, ordering, and managing inventories of expensive (but prone to obsolescence) materials that went into most of the company's products.

In each case, the projects were sponsored by a single senior executive with primary involvement in the business process. The senior executive team supported the executive's interests in getting the problems solved, but the executive peers' involvement ended with agreeing the project could go forward and supplying one or two people to the project team. All of these projects cut across a number of functions and departments in the company. One, for example, involved procurement, production and inventory control, product engineering, manufacturing engineering, marketing, sales, distributor relations, IT systems, product and marketing database, and others.

The typical structure of these projects was for representatives of each group involved, such as those listed above, to be assigned to the project full-time for six months or more. The project teams thoroughly researched the current state and produced a graphic representation of it; swim lane mapping was the most frequently used method. The team then produced a detailed description and map of an improved process. The improvement suggestions reached into virtually all of the organizational units involved in the process and represented on the team. The teams would then spend months developing briefing and presentation materials to present to the sponsor. Then, usually after major revisions requested by the sponsor, the team presented their findings and recommendations to the senior management team.

Not a single one of these projects succeeded in changing the process, other than changes the sponsor chose to make in his own area of responsibility. The other members of the management team had their own problems, their own point of view on how to improve them, and their own functional measures to worry about. They were not interested in changing how their organizations went about their business. The outcome was for the problem business processes to continue largely unchanged by the project. Indeed, several of these same problematic business processes later became Lean value stream process improvement projects for the internal Lean business process consulting team I led, 15 to 20 years after my involvement with the original failed conventional improvement projects.

Happily, the Lean value stream improvement projects were more effective than the early futile OD projects. The results of implemented improvements in the projects our team supported met the criterion of the 50 percent rule. On average, they reduced total lead time, process time, errors and rework, and handoffs by 50 percent in the first project, and there were similar improvements in the series of further projects that followed.

I attribute the success of the Lean value stream approach almost solely to the executive governance and accountability processes under which they operated. Senior executive support was important, especially when some of the first few projects encountered managers who failed to honor commitments they made to the future state implementation plan. Equally important was involving, from the outset, managers related to the target business process as members of each project's resource, governance, and accountability (RG&A) panels. In that role, they were invited and encouraged to engage during the workshops in vigorous give-and-take with the project teams. There they added to the discussion information they had but the team members lacked. They also learned from the grasp of reality the team members had but they as managers lacked. Altogether, most of these key managers bought into and benefitted from the improvements, and became repeat customers of the Lean value stream consulting team.

Organizational Governance for Enterprise Value Streams

Acting to improve an enterprise value stream is an organizationally and politically complex undertaking. Success depends on creating the conditions for the organization to recognize and support a cross-functional approach to improvement projects and business processes. Establishing these conditions does not mean changing organization structures or reporting relationships. At some point in its future an organization might want to take those steps, but they are not a place to begin (Figure 6.2).

Senior leadership needs to establish governance and management processes that support a cross-functional perspective in order for an enterprise value stream improvement activity to survive. The cross-functional perspective is necessary both for initial improvement projects and for ongoing process management after the transition from project mode. Key senior leaders, at the level of CEO and senior VP, can regularly act as a steering and oversight team for Lean initiatives. They can require Lean projects to have parallel steering groups—the directors or VPs of functions involved in the value stream targeted for improvement—with one person assigned as value stream sponsor. The senior leaders should expect the development of value stream, or business process, measures as one outcome of value stream improvement projects.

Value stream or process measures typically include worker safety (where it applies in enterprise processes), quality (first-pass yield or percentage of total lead time consumed by waiting, errors, and rework), speed (end-to-end lead time), productivity (process time, or actual hands-on time as a percent

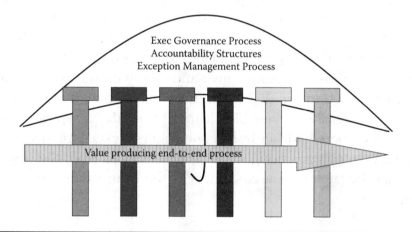

Figure 6.2 Senior executive support for end-to-end value stream process improvement projects.

of lead time), and externally incurred costs. These measures should show the projected improvement between the current state and proposed future state value stream design, and reflect the effectiveness of the future state implementation.

A particular characteristic of process measures makes them especially effective in sustaining the gains as improvement projects make the transition to the mode of ongoing business process (see Figure 6.3). Process measures are *scalable*. Being scalable means the measures apply to the business process as a whole and to specific responsibilities in each of the process functions. In an end-to-end view, the measures show the velocity, productivity, and freedom from errors and rework at the scale of, for example, the order management process, or month-end closing process, or new product launch process, or post-sales service process. Because the measures can be scaled down, they also apply at the intersections where the business process passes through the participating functions.

To illustrate how safety, quality, speed, and cost can be scaled to apply to functions, here are some examples of process measures applied to the intersection of value streams with functions:

◼ How quickly available and free from error is the scheduling information the order management representative relies on from master production control?

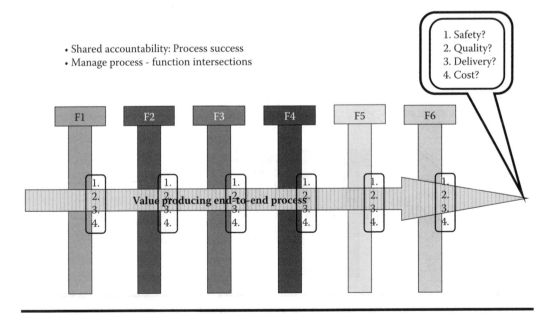

Figure 6.3   Functions commit to goals for function plus process performance.

■ How quickly are lab and imaging results available?
■ How accurate are the material availability and delivery transit times from the inbound and outbound logistics departments and the supply chain department?
■ How long do patients wait to be seen?
■ Are the pricing and terms information current as supplied by the contracts and pricing databases?

The measures at the process-function intersections provide objective information about the degree to which functions are meeting their commitments to the performance of key business processes. The senior leader oversight group can rely on these process measures to assess the functional VPs' and directors' accountability for agreed support of business processes in addition to their functional goals. And the measures form an objective basis for horizontal, peer-to-peer accountability among the project or process level oversight group.

The most important point is the organization's recognition of the joint performance responsibilities for functions as well as processes. This does not require a change in structure, but only that for the purposes of supporting the value stream business process perspective, the organization acts *as if* it were organized by process and has the governance, measurement, and accountability infrastructures to support its critical-for-the-customer enterprise business processes.

All of this may sound simple enough, but there are many moving, and potentially conflicting, parts to end-to-end value stream improvement projects. Experience with nearly 200 of these projects has shown the key for success is a set of well-defined and documented processes and a substantial dose of structure. Every one of these projects, by design, focused on a value stream that crossed at least one, and usually many more, organizational boundary, internal as well as external. The projects averaged 50 percent reductions in process time, cycle time, errors, and rework, and 50 percent improvements in percent complete and accurate information at internal and external points of handoff. The overall effects were uniformly substantial chunks of capacity freed up from previously tangled and inefficient cross-functional business processes.

Accountability in the context of these projects means the project leader, usually an internal lead resource, follows the process and its sequence. Figure 6.4 summarizes the steps in a value stream project; Table 6.3 provides details on who is involved, when, and what their involvement entails.

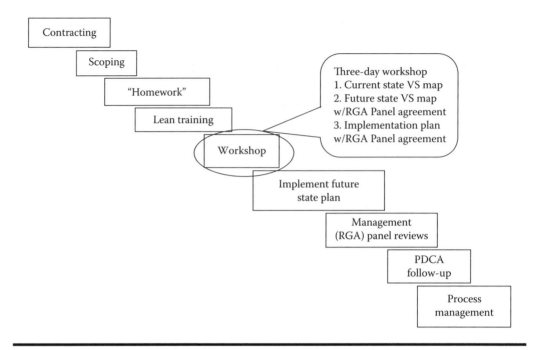

Figure 6.4   Steps and flow for value stream improvement projects.

Of equal importance, it is the responsibility of the project leader that all involved meet the commitments they have made to the project. This can be uncomfortable for the project leader in the first few projects since he or she is usually several levels below the managers, directors, and VPs on the project's resource, governance, and accountability (RG&A) panel, and whose frontline people are on the project team. These managers may not realize that *this* time, in *these* projects, the organization is committed to following through and implementing the actions they agreed to with the project team.

The project leader initiates the follow-through, coaching the manager/director/VP about meeting the commitments he or she made to the project team. It usually takes no more than two or three project sponsors being invited to a senior executive steering team session to review the current red status of the project's A3 project plan for accountability to project commitments to become the norm. Word spreads quickly when a senior executive steering team member (often the sponsor's senior executive leader) asks the sponsor (and by extension, the sponsor's peers involved with the project) for a recovery plan to convert the red elements to green. Most managers quickly come to understand they will be accountable to commitments they make to these kinds of projects without needing further encouragement from the steering team.

Table 6.3   Overview of Steps: End-to-End Enterprise Value Stream Improvement Projects

Step	When	Who	What
Contracting	Before committing to a project	Lean facilitator(s), process sponsor	Define problem, set objectives, define beginning and end of the target process, identify other execs involved, and identify internal project manager.
Scoping	After completed contracting	Lean facilitator(s), project sponsor, other involved execs (RG&A panel members), project manager	Review project process, structure, problem statement, and objectives. Complete SIPOC (suppliers, inputs, process, outputs, and customers) scoping document, identify workshop team members, and set dates.
Homework	Before workshop	Facilitator(s), project manager, workshop participants individually	Interview (one hour) each team member at his/her workstation regarding his/her work in the value stream, overview the project process and nature of his/her role on the project, observe how he/she does his/her work, and answer his/her questions.
Lean training	Shortly before workshop	Facilitator(s), project manager, team members	Two-hour overview Lean training, including hands-on practice value stream mapping exercise.
Workshop	Soon after training	Facilitator, project manager, team members	Day 1: Create current state map identifying all the problems, and prep team for breakthrough future state.

Continued

Table 6.3 (*Continued*)   Overview of Steps: End-to-End Enterprise Value Stream Improvement Projects

Step	When	Who	What
			Day 2: Team creates breakthrough future state map (with coaching as needed). Review current and future state maps with RG&A panel, and reach agreement on future state direction.
			Day 3: Document kaizens to implement future state. Review implementation plan, and reach agreement with RG&A panel.
Implement future state	In 60- to 90-day phases	Kaizen leaders from workshop plus kaizen members recruited as needed; weekly status reviews: kaizen leaders with project manager	Kaizens documented on A3 project plans with weekly milestones. If green, what's planned next? Lessons learned? If red, what's recovery plan? Help needed? Lessons learned?
RG&A reviews	At three- or four-week intervals	RG&A members, project manager, kaizen leaders, Lean facilitator(s) (optional); face-to-face or conference call	Restate problem statement and objectives of each kaizen, report kaizen green/red status, discuss next steps or help as needed. Remind all of commitments made to project implementation plan.
PDCA (plan, do, check, act) follow-up	Lean facilitator(s), project manager, process sponsor	90 days following project completion	Working as planned? Other needed improvements, opportunities uncovered? Responses needed to problems still unresolved?

Table 6.3 (*Continued*)   Overview of Steps: End-to-End Enterprise Value Stream Improvement Projects

Step	When	Who	What
Process management	RG&A panel and Lean facilitator	90-day intervals	Review process measures for value stream (safety, quality, delivery, cost, morale). Performing as planned? Identify further improvement opportunities, and sponsor a follow-up value stream project or other improvement projects as needed.

Lean Leader: Point of Contact with Executive Team

The organization's internal Lean leader acts as the senior executive team's point of contact for the enterprise Lean initiative. It is important that the senior team and Lean leader trust each other and have open lines of communication. The Lean leader should be familiar with the state of each project and provide regular (monthly, for example) summary updates of status and project outcome measures. In this and other capacities, the Lean leader acts as secretary to the senior steering team, able to interpret its direction and make priority rankings of which projects to take on in what sequence.

Familiarity with the status of all current projects and his or her role with the senior team allows the Lean leader to invite project sponsors to update the senior team when a project has stalled beyond the ability of the Lean leader to get it moving. It also allows the Lean leader to invite sponsors and project teams to update the steering team as a form of recognition. Once the enterprise Lean initiative is well underway, senior executive steering team meetings should become less frequent, even to the extent of being replaced entirely, instead using the time for one-on-one gemba walks in Lean project areas with an internal Lean resource. (See Chapter 8 for a discussion of gemba walking executives.)

Process and Structure

Any workshop-based value stream improvement process shares a common visible element, the value stream workshop itself. What distinguishes this model is the emphasis on its less visible elements. These elements are the process, structure, and cadence, particularly in the implementation phase after the workshop sessions.

Establishing and following a defined process (such as in Figure 6.4 and Table 6.3) is critically necessary for project success. Project sponsors and their peers often want to shortcut the process, for example, of carefully defining in advance a problem statement and objectives, or in defining in advance what is in and out of scope in a project. Or, because they typically are unfamiliar with which internal and external units are involved with the value stream, they leave critical participants off the list of those to be involved in the project.

Your job is not to provide the future state value stream map for the team. Instead, your job is to remind the team that they have been asked to make breakthrough improvements, not just tidying up around the edges of a problematic process that frustrates customers and workers alike. As they prepare to create the future state map, remind them of the 50 percent rule as a criterion for their work: reduce lead time by 50 percent, reduce wait time by 50 percent, reduce process time by 50 percent, reduce errors and rework by 50 percent, and reduce handoffs by 50 percent.

RG&A Workshop Participation

Remind the RG&A panel members of the same breakthrough expectations before they join the team to briefly review the current state map. Bear in mind that none of the RG&A panel members will have any prior grasp of the end-to-end current state process as it actually functions. They have been expected to and rewarded for concentrating on their functional areas. Remind all of the expectation for a vigorous top-down–bottom-up discussion as the team and panel grapple with the rationale for and implications of the proposed future state. Then dive into the future state map. Make sure at the end of this discussion there is explicit agreement on the future state direction, including any modifications arising from the discussion.

The third day in this workshop model produces the implementation plan for the future state, or at least the plan for the first 60 to 90 days of the implementation. Time and pace are important in Lean. Segmenting implementation into these time-limited windows acts to pace the kaizens identified on the future state map. It avoids the wasted effort of planning too far ahead, and leaves room to incorporate lessons learned in the initial implementation work in the following phase of implementation. You may need to coach team members who volunteer to lead the kaizens as they write the objectives, measures, and step-by-step, dated milestones to complete the kaizen in the 60- to 90-day window.

The dual-structured reviews during the implementation phase distinguish this model. The first review cycle is weekly, a brief check-in by project manager with the kaizen leaders. Referring to the Gantt chart on the one-page A3 project plan (see Figure 10.1 in Chapter 10) for the kaizen, the project manager can quickly see and ask the key questions:

- Is the planned work on schedule?
- If a milestone is due, is its status green or red?
- If red, what was the cause, and what is the recovery plan and expected completion date?
- Does the kaizen leader need help or support?
- What has he or she learned?
- What does the leader expect to accomplish in the next week?

The weekly review cycle is a constant reminder of the commitment the workshop team and kaizen leaders have made to the project and the RG&A panel. The weekly check-in also means the kaizen cannot be more than a week behind schedule before the project manager knows about it and can intervene, or call on the Lean facilitator, an RG&A member, or others as needed to get the kaizen moving again.

RG&A Reviews Two-Way Commitments

The second review cycle is monthly for a 90-day implementation plan, or at three-week intervals for a 60-day implementation plan. In either case, the RG&A reviews can be held via conference call with the kaizen A3s distributed in advance. This review cycle reminds the RG&A panel members of commitments they made to the kaizen teams and to the project and, by implication, to the senior executive steering team. And, it reminds the kaizen teams of commitments they made to the RG&A panel. The RG&A reviews keep the panel members up to date on progress and accessible to the kaizen leaders, able to hear requests for or to offer help or support.

These reviews are also based on the one-page A3 project plan for each kaizen. No further reporting material is required, no slide show to put together or sit through. The kaizen leader reads the A3's problem statement and objectives, reports the green or red status, describes problems encountered, makes requests, if any, relates lessons learned to date, answers questions, and is done. As noted earlier, if a project sponsor is asked to update the senior steering team, the project's A3 with a Gantt chart line for each kaizen, coded green or red, is the only document needed or expected.

Your job as a Lean facilitator is not to be talked into shortcuts. In the many projects where I have been involved, the only unsuccessful ones were the three in which we failed to follow the process. Trust the process! Work through the details of your role with the senior team, including bringing the sponsors of faltering or stalled projects to them for review and support. Make sure the project sponsor understands the commitment to be informed about the current state by those who actually do the work in it, i.e., the workshop participants and the maps they create. The sponsor's preconceived solutions might turn out to be off target. Invariably, early "solutions" need to be revised, whether yours, team members', the sponsor's, or RG&A panel members'; nobody is familiar with the process beginning to end. The project is an exercise in grasping reality for all involved, including middle- and higher-level managers.

Do your homework! Meet the workshop participants at their workstations. Ask them to tell you about their part in the process, to show you their work and how they go about it. Tell them what to expect in the workshop, assure them they are already prepared based on their knowledge and experience with their day-to-day work. If they have never been involved in Lean before, let them know the preworkshop training is to prepare them with Lean's basic concepts and with practice in value stream mapping. Answer their questions. Absorb what they tell you about their part of the process as part of your preparation and to minimize being surprised in the workshop.

Summary

Lean principles and tools are effective in enterprise business processes just as in manufacturing. And, just as in Lean manufacturing, sustaining enterprise Lean applications requires a parallel implementation of Lean management. Lean tools apply in office processes just as effectively as in manufacturing, from 5S and workplace organization, work paced by takt time, work cells, in-cycle work separated from out-cycle tasks, defined flow paths, and mixed-model production with controlled release through heijunka to level mix and volume. However, until frontline workers in enterprise processes experience the benefits of Lean, they can be especially sensitive to the increased accountability that comes with standards for performance and productivity, and with the use of visual controls.

Enterprise value streams usually lack the singular ownership equivalent to that of a plant manager in a factory setting. That, combined with crossing so many more internal organization boundaries, makes leadership alignment an even greater issue, and therefore a greater focus in Lean enterprise implementations. Even considering the greater political complexity of enterprise Lean, the gains in performance make the added effort absolutely worth it. Improvements in performance from successful enterprise Lean initiatives produce at least as much benefit as Lean manufacturing initiatives.

Study Questions

1. Has your organization applied Lean in administrative, technical-professional, or service delivery processes? Has the experience been successful? Why?
2. Has your organization used value stream maps of business processes as part of your improvement activities? Have the maps been helpful? Why?
3. Does it seem reasonable to think that 50 percent of the work time spent in enterprise business processes is non-value-added activity? Why or why not?
4. Have past business process improvement projects been successful here? Why or why not?
5. Have senior leaders been involved in business process improvement projects other than "kicking them off"? What effect has this had on the ultimate success of these projects?
6. Has there been evidence of accountability to commitments made to project teams working on improvement initiatives? What has been the effect on the ultimate effectiveness of the projects?
7. What, if anything, in this chapter did you find surprising?

LEARNING LEAN MANAGEMENT AND PRODUCTION: SUPPORTING ELEMENTS OF LEAN MANAGEMENT

II

Chapter 7

Learning Lean Management: The Sensei and Gemba Walks

The principal elements of the Lean management system do not seem to be complex or difficult to understand. Yet they represent the rocks on which so many Lean initiatives have run aground. Why?

Part of the answer involves discipline, certainly. But, another part of the answer involves a teaching and learning model for Lean and Lean management, and how to be a Lean manager.

Lean's classic master-apprentice teaching and learning model has long proved effective with leaders in tactical positions, those responsible for and familiar with day-to-day operations (and Lean leaders responsible for supporting Lean implementation). Or, it is useful to think of the tactical chain of command as extending from a plant manager or a customer service director, or a director responsible for surgical operations down through floor-level team leaders and Lean resources. This chapter covers the approach for learning Lean and Lean management for those new to Lean in this segment of leaders, and in some cases for their bosses as well.

Before continuing, however, it is important to make a distinction I have learned to recognize in the ten years of experience since the first publication of this book. What works for those directly involved with tactical day-to-day operations is not the same as the approach that has proved to be effective with senior executive leaders whose concerns are usually more strategic in nature (Case Study 7.1). Find that approach, case studies, and tools in Chapter 8.

CASE STUDY 7.1: GEMBA WALKING A VP
FOR LEAN PRODUCTION

A new VP had just been transferred from marketing to take over manu-facturing. His undergraduate degree was in engineering in addition to an MBA. He had not worked in engineering or operations for many years. The three-person team responsible for the Lean initiative across the 12 manufacturing plants met with him his first day on the job. He told us to "treat him as a student." We gave him several books to read and asked him to gemba walk with us individually, one on one. He agreed.

My colleagues and I gemba walked the factory floors with him, stop-ping to ask him to observe something and asking him, for example, if he could tell from a generic reorder card the specific quantity of parts to be produced. Or, how he could know amid the mountains of parts crowding the area whether parts needed to be reordered, and what information would be needed to answer the question. He was a quick study, able to answer any of our questions after a brief pause.

As he learned from us, he began asking similar questions of his plant managers on frequent drop-in visits to gemba walk in the plants. The plant managers scrambled to get up to speed with Lean, suddenly becoming much more interested in our help.

Five years later, now as a corporate officer and senior VP for global operations, he was the key sponsor on the senior executive team for an enterprise Lean initiative at a critical point for the business. Despite the objective success of enterprise Lean, his peers, VPs in other areas of the business, lacked interest in our gemba walks focused on the nitty-gritty details of Lean applications in enterprise administrative operations. The initiative was losing support and was on increasingly shaky ground until we developed a way to engage our senior executive clients. See Chapter 8 for the details of our recovery.

In the master-apprentice model, the master (the Lean sensei, or teacher) and apprentice walk the floor, a gemba walk.* The master points out examples where a Lean approach might apply and asks probing questions to stimulate the appren-tice's thinking. The master continues until she is satisfied the apprentice has

---

* This model may be best known from the practice Toyota used for Lean training through the 1990s, at which point attrition in its Lean sensei ranks and its rapid growth led to more reliance on con-ventional training methods.

grasped the concept (for example, produce smaller quantities more frequently, replenish only what has been consumed, make sure what is needed for the work is readily available at hand and obvious at a glance, arrange the work so as to minimize or eliminate wasted motion, and so on). Then, the master makes an assignment, usually in the form of a suggestion. "Why don't you see if you can apply this concept here? We'll see how it looks next time" (often, next week).

Your Sensei and "True North" Provide Direction

Lean is tricky, because it is so much more difficult than it seems. Reading about Lean, attending workshops on Lean, or participating in simulations is no substitute for real-world, on-the-floor coaching and critique from someone who has experienced implementing Lean. An effective sensei will be insistent and at times critical, to the point of being at least mildly abrasive. Most of us need a sensei to help us understand how to translate basic Lean concepts into actual functioning applications. More than that, your sensei should also instill in his or her students the discipline needed to effectively sustain a Lean conversion and have it improve consistently.

Toyota refers to "true north" as its ultimate direction for perfection in Lean. It has relied on its internal sensei to keep the company on that path, just like early travelers relied on the North Star, Polaris, to keep them headed in the right direction as they traversed unfamiliar territory. In a Lean implementation, the sensei plays the role of Polaris. Working with a sensei is not like conventional training. Instead, it more closely resembles an internship or apprenticeship. In these models of teaching and learning, the student learns over time, through experience from applying concepts to actual live situations and carefully observing the consequences under the tutelage of the sensei.

The classic sensei is Socratic in approach, teaching by asking questions that stretch the student's thinking and perceptions and stimulate the student to consider entirely new possibilities. Students might encounter questions such as:

■ How can you tell what is normal in this area?
■ What would you learn if you measured in smaller intervals of time?
■ Can you tell what is supposed to be here—and what should not?
■ How frequently are the supervisor and his or her boss in this area? How can you tell?
■ What is the team leader supposed to be doing in this situation?
■ Why should you expect the team leader to know that?

- How could these expectations be made clearer?
- How do you know that the designated person carried out these posted procedures?
- How could someone know who was responsible for this task?
- How could you know these things with more certainty?

Some of the sensei's teaching is likely to be typical classroom instruction, especially early on when introducing the basic concepts of Lean or of a particular technique or approach. To be effective, these sessions should be immediately translated to observation and application on the production floor, a destination to which the sensei will often appear impatient to return. Once the conventional training is over, sensei and student begin or resume gemba walking to reinforce what has been presented and to extend the lesson through extending the principles to situations encountered where people are doing actual work (Case Study 7.2).

Think of your sensei as a personal trainer, one who sets expectations for you and then teaches, coaches, and prods you to meet them. Can you learn Lean production and Lean management by yourself through study, application, and self-critique? Certainly, you can. Perhaps more pertinent, however, is how long do you have? How persuasive can you be with your peers and superiors as you work your way through the learning curve, stumbling occasionally as anyone would with brand new ideas? In most cases, probably not long enough. This is a place where calling on outside expertise is well worth it, subject to all the caveats of working with external—or internal—consultants, as in Case Study 7.3.

CASE STUDY 7.2: LEARNING TO SEE PROBLEMS, NOT JUST CORRECTING THEM

In one case, I was working with a supervisor on basic 5S discipline in a welding department. Except for the weld tips themselves, nothing in the place had been cleaned since it opened many years earlier. You can imagine its condition: grime and dirt seeming to grow from every surface—horizontal and vertical, top and bottom. After a few weeks of pointing out several particularly dirty areas and talking about the benefits of discipline and order, it became clear that the supervisor was only cleaning up the specific spots we had looked at. I admonished him: "Learn to see the dirt yourself!" He did, took initiative to get the entire area cleaned up, and needed no more coaching on this topic.

CASE STUDY 7.3: DON'T USE CONSULTANTS AS A PAIR OF HANDS

A regional health services organization became convinced Lean was the way it was going to survive and prosper in its rapidly changing industry. It had no internal expertise with Lean concepts or their applications in healthcare settings. To begin its Lean journey, the organization engaged a consulting firm whose approach was to implement Lean through a series of kaizens. For two years, every time a kaizen was identified, consultants from the firm came in from out of town to lead the kaizens. In all that time, they did not train anyone in the client organization to lead or facilitate a kaizen event. The client eventually stopped working with the consulting firm and developed its own Lean resources. Since then, it has gone on to develop deeply rooted and widespread Lean thinking and a robust and effective internal Lean application capability that is a model for many in the industry.

Another healthcare firm, consisting of a mutual health insurance business and a number of physician practices, or clinics, in a large metropolitan area had heard about Lean successes in healthcare. It too was convinced that Lean would help it survive and prosper, in this case in the rapidly changing environment as the 2010 Affordable Healthcare Act moved toward implementation.

The firm identified roughly a dozen internal resources to be trained in Lean tools and Lean implementation, and used outside experts as trainers. Once training was complete, the internal resources were deployed across the organization, each one linked with one to three specific departments. Three years later, most of the internal resources had their employment terminated. A few had returned to their previous positions.

When asked about the reason for this abrupt change of direction, the VP responsible for process improvement indicated that only two of the Lean staff had developed into Lean coaches and teachers; the rest of the internal consultants acted as a standby fix-it staff. When a department had relatively minor internal process problems, the standby fix-it resources were given the problem to own and fix. In the course of their three years on the job, only 2 of the 12 Lean consultants had taught anybody Lean thinking, not to mention bringing Lean applications to problems of significance beyond any single department. Most of the staff had simply been a convenient remedy for minor departmental irritations rather than resources enabling strategically important improvements. When budgets had to be cut, almost nobody but their own VP spoke up for the potential strategic value of the fix-it resources. True, they had been a convenience for the departments, but they had not demonstrated anything their internal clients considered essential or that helped to invigorate a stalled corporate strategy for improving effectiveness and efficiency.

In one important way, the sensei should be like any good consultant, recognizing the division of responsibility between client and professional advisor. That is, the sensei or consultant is responsible for teaching, giving advice, stimulating new thinking, and identifying new directions. The client is *always* responsible for decision making, that is, for whether and how to apply the advice of the teacher. In this relationship, the only decision the sensei makes is whether to continue working with the student, based on how the student follows through on commitments.

Gemba Walking

Gemba is roughly translated from the Japanese as "the real place." In this sense, *real* refers to where the action is happening. To illustrate, Japanese television news reporters covering the devastating 2011 earthquake, tsunami, and nuclear plant disaster that struck Fukushima introduced themselves on camera standing in front of a site near the disaster zone as "reporting live from gemba in Fukushima." If your focus were on improving customer service in a call center, gemba would be the call center floor and workstations; for a hospital's emergency services, gemba might include the ER waiting room, registration stations, and patient care areas. For manufacturing, gemba is the production floor.

The idea of gemba is simple:

1. Go to the place.
2. Observe the process.
3. Talk with the people.

Gemba walks typically take place on a regularly scheduled basis. An optimal schedule is at one-week intervals. This is enough time to allow the student to digest the lesson and complete assignments the sensei gives, but short enough to maintain a sense of pace. In some cases, the assignment will be to develop an explanation for why something appears as it does and what an alternative might be. As the teaching and learning progress, the gemba walks shift to the sensei suggesting assignments for the student, and then following up on the next walk. The gemba walks are another instance of accountability for bringing actual (what the student has been able to accomplish) in line with expected (the assignment from the sensei).

In these cases, gemba walks become a method for setting and following up on expectations for Lean learning and implementation. As the student begins to correctly interpret what the sensei points out, the sensei will make assignments to implement the concept the student has seen on the gemba walk. Next, the sensei will expect the apprentice to initiate similar action when encountering a need for applying the same concept, but in circumstances and processes different from those in the areas where they have already walked (Case Study 7.4). What is being tested

---

CASE STUDY 7.4: HOW GEMBA WALKS CAN REVEAL OPPORTUNITIES FOR IMPROVEMENT

Illustrating this aspect of gemba is a case where a value stream manager had brought his area from chaos to stability and excellent performance. The area missed its daily production targets and pitch rhythm only when materials failed to arrive on time from outside suppliers. The leader was restless; he believed the area could improve further, but did not know what to do. We went on a gemba walk, looking for opportunity.

From the day the plant opened, parts were unloaded from the paint line into kit racks that held components for ten units. The units could be up to 6 feet long and 4 feet wide, so the racks had to be big enough to hold the components safely. They were big, about 8 feet long by 4 feet high, heavy, and not easily maneuvered. Formerly, they were staged at two-person assembly benches where each unit was built complete from start to finish by the two-person team. Now, they fed an L-shaped assembly line. The painted parts held in the kit racks were hung on the paint line in groups of four units.

I asked the leader if the balance on his assembly process would improve if it were arranged in the classic U shape. "I can't do that; the racks wouldn't fit inside a U," he said, and then quickly realized the implications of cutting down the racks for tightening the footprint of his assembly line, bringing better balance to it, and improving its performance.

here is the apprentice's ability to recognize the concept and its potential application in unfamiliar areas, processes, and settings.

How Lean Typically Starts and Grows

The orthodox or purist approach for bringing Lean to an organization calls for the Lean sensei to start at the top. In this view, the Lean sensei starts by gemba walking the CEO to teach him or her to see through Lean eyes. The sensei then works his or her way down through the organization structure, building knowledge and support for Lean from the top down.

In my experience, this approach is unrealistic.

In 15 years I have seen the top-down sensei approach work in a large, complex organization only once. On rare occasions I have seen it work in small manufacturing companies or self-contained administrative operations units. But in the vast majority of cases, senior executives, even in smaller organizations, do not have the time, patience, or interest in learning Lean to the depth and in the way the sensei learned. They have neither the patience nor willingness to learn in the way the sensei wants to teach. And, they typically are not interested in learning to become a Lean implementer; they have people like us to worry about implementing the nuts and bolts of Lean tools.

Chapter 8 addresses why most executives respond this way, and then describes an approach, process, and tools that have proved effective in engaging senior executives with a Lean initiative via learning the Lean management system. The rest of this chapter covers gemba walking and process focus in an operations environment, in the best case beginning with the senior operations executive, as in Case Study 7.1.

By far the most typical scenario I have seen is for Lean to start in the middle of an organization, then spread like a slowly propagating beneficial virus. In this scenario, one or a few people somewhere in the middle of the organization, say, an engineer or department manager, have gotten the Lean bug. They have read something, been to a conference, visited or toured a Lean operation somewhere. They begin experimenting with small-scale Lean applications within their sphere of control. Their success piques the interest of others and the experiments expand. As Lean applications succeed and spread, the improved results catch the attention of their superiors. Eventually, the organization hires a consultant to help Lean grow, or forms an internal Lean resource team for the same purpose. At that point, the Lean initiative has access to people in more senior operations positions. Now gemba walks can begin higher up in the operations chain of command, either in response to a request from operations executives or in response to a proposition from the Lean advocates to gemba walk their operations superiors (as in Case Study 7.1).

Gemba Walking Teaches How to See in New Ways

People in hierarchical organizations (which are, after all, most of us) respond to the requests, suggestions, and directives of their superiors. As the superior learns to ask for and teach how to apply the principles of Lean production and Lean management, the subordinate is likely to listen carefully, learn the new expectations, and learn how to comply.

Gemba walking is a practice with a definite tie to the expected versus actual theme in the Lean management system. This explains the Lean sensei's desire to start his or her regimen of gemba walks as high in the operations chain of command as possible. When those in positions of greater authority learn to point out forms of waste on the operations floor that have escaped the notice of, for example, the plant or unit manager, it creates powerful motivation to get up to speed with what the boss expects (see Case Study 7.1).

And when superiors are able to teach Lean principles, expect to see them applied, and are able to evaluate progress on sight, the chances for sustaining a Lean initiative greatly improve from Lean's informal beginnings somewhere lower in the organization.

The object of gemba walking is teaching to see through different eyes what the student has been looking at for an entire career. The alternative is for the student simply to follow the directions of the teacher to "do this" and "do that" without being taught or understanding how the assignment relates to a Lean rationale or principle. Even worse is when a consultant, internal or external, selected to teach and build internal Lean capability, instead simply does the work himself or herself, acting as a "pair of hands," as in Case Study 7.3. Some, but usually not much, learning happens this way, and when the sensei departs he or she leaves no lasting transfer of knowledge.

The desired outcome is for the student to learn to see where the principles of Lean or Lean management can be applied. Importantly, this is because application of the principles is an invention derived to fit the unique nature of a given situation. The student's knowledge is best demonstrated by his or her ability to see the application of the same principle in completely different environments and work processes. Where possible, the teacher will want to see the student transfer what he or she learned in a factory to an administrative process, from a healthcare or office setting to a physical production process, or at least in different ends of processing areas in the same setting, or in an entirely different setting, be it manufacturing, administrative, healthcare, or services.

Seeing through the Surface

This is the skill of being able to "see through" the surface of any given work process to the underlying potential application of one of the handful of Lean principles. For example, a clerk in finance accumulates a week's worth of credit card charges and then posts them all at once. This, he or she tells a gemba walker, is more efficient for her. An operator of automated equipment with an order for 5,000 of one part number produces 35,000 because, he or she tells a gemba walker, the job was running well and he or she figured letting it run was more efficient than stopping to change the material and tooling to the next part number on the machine's order board.

What is the same in these two examples? Is this batching? Is one or both an example of the waste of overproduction? What is the effect of each on the value stream? On each facility? Does the inventory produced or accumulated add value or add cost? What would be an alternative approach and method, and what should be points of focus in developing the alternatives? These are the kinds of situations that clearly test a student's or a practitioner's grasp of Lean concepts.

Few Lean applications are literally answers to be found in a book—including this one! As Lean or Lean management apprentices become more skilled through gemba walks with their sensei, they gradually develop their own expertise as gemba walkers, teachers, and assessors. It then becomes their turn to gemba walk with their subordinates or others, teaching and helping them develop their own mastery of Lean production and Lean management.

Learning to be a Lean implementer and teacher through gemba walking requires patience and high tolerance for frustration. It is not fast. One of the main ways of learning is from experience as you work to make the corrections and refinements you find necessary (often prompted by your sensei) as you apply newly learned concepts and tools. There is no good alternative to gemba walks as the method to learn Lean production and Lean management. That is because Lean is a mindset, a different way of seeing. Learning to see through different eyes is a process, one that develops over time with practice, feedback, and hands-on experience (Case Study 7.5). If that is the bad news, the good news is gemba walks are extremely effective as

CASE STUDY 7.5:   LEARNING LEAN MEANS EXPERIENTIAL LEARNING

A manufacturing VP who had a good grasp of Lean concepts agreed to participate in a kaizen event, the goal of which was to simplify the process in a workstation where component parts were subassembled before going through a finishing process and then to final assembly. The VP's Lean preparation had been by gemba walks with internal Lean resources, in which he learned to see where Lean tools could apply. He was a quick study, and had soon begun gemba walking his subordinates, energizing the Lean initiative throughout the company's manufacturing operations.

The kaizen focused on ergonomic and efficiency improvements, including reducing the operator's walking and reaching in the workstation. One element of the improvement addressed the way the parts to be subassembled were presented to the operator. One of the components was a steel stamping, almost a foot long, with several complex folds and protrusions. All the parts were delivered to the workstation in separate plastic totes. The long stamped parts came tangled together in a larger tote. Each time the operator reached for a part, he or she had to untangle a piece as if solving a puzzle. It was time-consuming and clearly seemed to involve wasted effort, a burden on the operator.

Among the ideas in the kaizen was the VP's observation that if the parts were placed in the totes in a nested arrangement, one against the other, they could be in a smaller tote for less reaching, and easily removed one at a time. It would save several seconds each cycle and improve the ergonomics for the operator.

Some of the kaizen team members, including the VP, tried out this idea. They arranged several totes of these parts in nested order. Sure enough, they proved easy to remove one at a time. They showed this to the operator and asked her to try it. It turned out the team had failed to account for the fact that because the stampings had sharp edges, operators who handled them had to wear gloves, a safety requirements for that station. Wearing the proper gloves, the nested parts fit together too closely to be easily picked, requiring almost the same amount of jiggling and jostling as the original presentation.

> The team went on to make other improvements in the ergonomics and flow of work through the workstation, but nested parts was not among them. Lean concepts are easy to grasp in the abstract, but it takes practice to learn to see how they might apply. Ideas for improvement can be easy to see, but attempting to apply them in the real world often teaches experiential lessons that are not so easily seen.

a learning model. Through gemba walking you gradually establish a new, durable, Lean way of seeing and thinking. Six months of weekly gemba walks is on the low end of the period necessary to develop eyes for Lean and a Lean approach to managing.

Focus on Both Processes—Technical and Management

Focusing on process is essential for success in a Lean operation. That focus must include the elements of Lean production and the elements of Lean management. The components of these two aspects of Lean—technical production and management systems—are often intertwined with each other. That means effective Lean leaders need to be well versed in both sides of Lean.

Table 7.1 shows some aspects of Lean management that a gemba walk might focus on. Each item is discussed in detail throughout this book. The point here is that a gemba walk can focus on far more than the technical aspects of Lean production. When you know what to look for, evidence for the presence or absence of a robust Lean management system is everywhere on a production, processing, or service delivery floor. Gemba walking, traditionally focused on the technical side of Lean (is inventory being pushed or pulled? is standard work balanced to takt?), is indispensable in learning Lean management and especially in maintaining it.

As leaders become proficient in the first several levels of Lean, gemba walks are still useful for asking operations leaders to reflect on what they have done and learned, and for challenging them to go further. This is particularly applicable in situations where leaders have led progress to a plateau and either do not know how to push further improvement (as in Case Study 7.4) or have become satisfied with the current state. Here, gemba walks are likely to encounter the bedrock of the batch production mindset, the old-school view: "If it ain't broke, don't fix it." This runs counter to the Lean principle in which perfection is the goal, and the related tenet that everyone in a Lean system has two responsibilities: to run the business and to improve the business.

Table 7.1 Looking for Lean Management

What You Should See	*What People Should Know*
Process Focus	
• Tracking charts show current actual versus expected status for all processes, in-cycle and out-of-cycle. • Production-tracking charts initialed by supervisors at least twice daily. • Reasons for misses noted on tracking charts.	• How are you doing at hitting your production goals? • How can you tell if out-of-cycle and daily or weekly tasks are getting done as they should? • (Leaders) Is there a regular schedule for gemba walks in this area? What is it? What happens on a typical gemba walk? • Have these charts resulted in process or performance improvements? Examples?
Process Improvement	
• Top three to five reasons for misses documented and visible at cell/line, department, and value stream information boards. • Summary project plans (A3s) for improvement posted and current at department and value stream information boards. • Employee suggestion system shows recent suggestions, current action on suggestions, and implemented suggestions with trend chart of numbers submitted and implemented. • A visual daily task assignment and accountability process is in use and current.	• What are the three biggest problems in this area? • How do you know these are the biggest problems? • Is any work being done on these problems? How can you tell? • Is there a regular method for operators to make suggestions for process improvements? What is it? Does it work? Examples? • How can you tell your suggestions are listened to? • (Leaders) What improvement activity is going on in this department? • (Leaders) How do daily task assignments work here? Is there regular follow-up on assignments?

Continued

Table 7.1 (*Continued*)   Looking for Lean Management

What You Should See	*What People Should Know*
Leader Availability	
• Team leaders on the floor in their process area virtually all the time and available to operators. • Supervisors on the floor in their process area. • Response system to summon supervisors, team leader, others when needed.	• (Leaders) How many hours/day on average do you spend on the floor? • How do you contact your team leader when you need him or her right away? • How quickly is help available when the process is interrupted by a problem the team leader cannot fix?
Labor Planning (at team boards)	
• Rotation path and starting assignments displayed. • Expected attendance chart up to date, displayed. • Qualification matrix up to date, displayed (including qualified out-of-zone operators).	• How can you tell who's supposed to be here on any given day? How can you tell when you have call-ins? • (Leaders) What do you do when there are call-ins? • How do you know how many people you need for a given rate of production? • Do you rotate jobs here? How do you know where you'll be working at the start of any given day? • How can you tell who's qualified to do which jobs in the area?
Standard Work	
• Operators and leaders have and are following their respective standardized work. • Standard work charts, complete with cycle times for in- and out-cycle work, are posted and clearly visible from operator workstations. • Leaders' standard work is displayed day by day for up to a week.	• Can you show me the standardized work or procedures for this station or role? Do people in this area follow standardized work? Does anyone ever monitor to see it's being followed? • (Leaders) What's your process for monitoring standardized work? How often do you monitor it? • (Leaders) Do you use standard work? Let's look at it for today.

Table 7.1 (*Continued*)   Looking for Lean Management

What You Should See	What People Should Know
Communication	
• Daily shift meeting agenda visible on the team info center. • Where applicable, info from other shifts is displayed in cell/line or department info board. • Team leaders', supervisors', value stream daily meetings occur.	• How often does your team meet as a group? Is it a regularly scheduled meeting or just once in a while? • (Leaders) How do you know what topics you'll cover in any given day's start-up meeting? • (Leaders) Do you lead or attend any daily meetings? What are they?
Workplace Organization	
• Weekly 5S audit form and action items for the week are current, displayed at team info boards. • Cleaning routines and checklists visible, current. • Total Productive Maintenance (TPM) checklists current at each asset. • Clearly visible indicators of location and quantity for each object in the area. • Signage or identified addresses for tools, WIP and raw materials, reorder points and max quantities, kanban cards. • No clutter, dirt, or debris on floors, shelves, tops of cabinets, under racks and conveyors, etc. • All horizontal surfaces clean. • Cabinets, drawers labeled, contents match labels.	• How do you keep track of housekeeping in this area? Are there standards for housekeeping? • (Of any object) What's this? How can you tell where it's supposed to be? How many of them should be here? • How much material are you supposed to have in this location? How can you tell? • What are the reorder points for (any and all) materials? What's the process for reordering?
Working Buzzer to Buzzer	
• Work starts and stops on time.	• What times are breaks in this area? Are people usually back on time or are there usually stragglers?

Being the Sensei: Gemba Walking as a Structured, Repeatable Process

A Lean sensei, if he or she is worthy of the name, should be able to make a Lean assessment in any area of your organization by following three basic steps:

1. Go to the place.
2. Observe the process.
3. Talk to the people.

He or she will likely ask you a bunch of questions, some of which you may not have thought of and to which you do not have a ready answer. Welcome to the sensei's process for stimulating your efforts to see through the surface and recognize application of Lean's principles! That approach by the sensei is probably fine if you are one of your organization's identified Lean geeks. You may well have volunteered or been asked to become an internal Lean resource capable of teaching others.

It is worth keeping in mind that you signed up for what is, in effect, this series of the sensei's seemingly random pop quizzes. Others—your students—might not take as well to such an unstructured routine. Remember, in your role as resource, coach, teacher, sensei—however you are identified—you are the supplier. Your students are your customers, and the first Lean principle is: value is defined from the point of view of the customer. Even so, it is perfectly appropriate to ask questions your students cannot yet answer. It is equally appropriate to suggest they think again about what they have learned, and how to apply it *right now, right here*. Their status as your customer does not entitle your students to expect you to think for them, to receive your answer before giving their own; that creates dependency rather than developing capability.

There is, however, an "on the other hand" here. That is, you could easily end up frustrating students whose learning style is not well adapted to what might appear to them as random walks and out-of-the-blue questions. Frustration does little to advance a student's knowledge and confidence. This might not be much of a concern among traditional hard-core Japanese sensei. Their gruff methods were supported by their employers, especially early on at Toyota. Or, working as outside consultants, Lean sensei risk only losing a client, rather than losing employment altogether. Neither of these conditions may completely apply to you, the internal coach.

Letting "Students" Know What to Expect

An approach to consider is to make your gemba walks a structured, repeatable process. The process is based on a Lean or Lean management assessment. A Lean management assessment—Lean management standards—can be found in Appendices A and B and as PDF downloads from dmannlean.com. The standards cover the Lean management system in manufacturing and nonmanufacturing operations. For Lean production, the publicly available Rapid Plant Assessment by Eugene Goodson is an example of an assessment for Lean manufacturing (www.bus.umich.edu/rpa; see also hbr.org/2002/05/read-a-plant-fast).

Using of one of these tools or something similar as the basis for gemba walking lets students know what process to expect even though each walk's content may differ. It eliminates surprises from the process and expectations differing from one gemba walk to the next. A set of standards provides them, and you, with a straightforward way to gauge their progress. (For the student: How well did I do assessing the status of the daily accountability process last gemba walk? For you: Is the student ready to move on to the next standard, or does he or she need another walk on this one?)

Each Lean management standard has the same one-page format. The standards assess eight dimensions of Lean management, each on a 5-point scale. As you begin the walk, you tell the student which element of Lean or Lean management will be the focus. You hand him or her the assessment sheet for that element. The one-page standard acts as the structure, process, content, and agenda for each gemba walk, as well as a picture for the student of what good practices look like.

The Lean management standards (Appendices A and B) include three features designed to be helpful in using them for teaching and learning. The first is the set of diagnostic questions at the top of each dimension's single page. A gemba walk for visual controls based on that standard could go like this: the teacher considers the student's progress on visual controls and what can be observed in an area for the gemba walk. Based on this, the teacher highlights two or three of the questions for the student to address on this gemba walk. It is the student's responsibility to make the observations and have the conversations needed to answer the highlighted questions.

For example, the standard for visual controls lists these diagnostic questions:

1. Can you see visual cycle or procedure tracking charts in the area? Do they show expected versus actual times?
2. Are the charts current to this or the last shift?

3. Are incidents that delay work described clearly (what we had but did not want, or wanted but did not have)?
4. Are visuals reviewed regularly? How frequently? How can you tell?
5. Can leaders and task-level people in the area cite improvements from problems noted on visual charts?
6. Are visuals used here for support tasks, e.g., materials, transport, attendance, assignments, qualifications?
7. Do leaders regularly review the visuals? How often? How can you tell?

The second feature is the self-describing levels on the assessment's 5-point rating scale, used on all the standards. The scale's 5 points are shown here:

1. Preimplementation
2. Beginning implementation
3. First recognizable state
4. System stabilizing
5. Sustainable system

The self-describing levels give some detail on what an observer might see in, for example, an area at level 2 (beginning implementation). The visual controls are shown here:

■ Some cycle tracking charts, irregularly filled in
■ Most charts record numbers, not document delays, problems
■ Where problems described, too vague for action
■ No or irregular review for action on problems
■ Visuals more "check the box" than tool to highlight problems, delays, and drive improvement

The self-describing rating levels provide some concrete examples for the student to consider in evaluating what he or she has observed about visuals on a gemba walk in a particular area.

The third feature is an immediate test and feedback. At the end of the gemba walk, as student and teacher leave the area, the teacher asks: "So, what rating would you give for the visuals in this area?" Student and teacher compare ratings. When they reach different conclusions, the discrepancy makes for an immediate teachable moment.

The teacher gives immediate feedback, indicating what he or she saw and why the observations supported the rating he or she gave. (Early on, students tend to give higher ratings than the teacher.) Each Lean management standard includes space for notes; student and teacher briefly compare what they recorded. The student gets immediate feedback on key points the teacher saw, and the difference between how the teacher and the student evaluated what they observed. Next time, the teacher will probably return to this dimension, visual controls, and at least one of the questions at the root of the difference between the student's and teacher's conclusions and ratings.

In these ways, a Lean production or Lean management standard provides the agenda, content, structure, process, and feedback in a consistent program of repeatable, predictable gemba walks.

Summary: Learning Lean Management by Being a Sensei's Apprentice

This chapter concentrated on two aspects of developing and maintaining focus on process. The first involves learning the principles of Lean by stepping into the role of student, engaging a Lean sensei to teach you the principles and guide you through initial applications. This will happen principally through your gemba walks with the sensei. Before you can teach others, you must develop an initial level of mastery of Lean concepts and learn to see where and how to apply them.

Second, you become the teacher through gemba walks with those who report to you, often as you continue to learn from your sensei. As Lean produces improvements and your ability to teach others develops, you will most likely develop the informal authority that allows you to be effective in coaching those who may be several organizational levels above you.

Keep in mind, the way your sensei taught you might not be effective with those who are new to Lean and less enthusiastic about it than you. If your teacher's approach to gemba walks appeared to be unpredictable and without a discernable structure, consider an approach where the ends are clear to the student, the process is predictable, and some criteria for good practice are spelled out. But do not hesitate to compare what you see in an area with what your student sees. The objective in gemba walking is for the student to develop eyes that see what the sensei sees.

It is not unusual for a Lean advocate somewhere in operations to drive implementation far enough to produce success that gets the attention and then support from senior leadership in operations. Engaging senior leaders from other functions in a Lean initiative that began in operations is the topic of Chapter 8.

Lean management is, as much as anything, a way of thinking. A paradox is that this way of thinking arises from new ways of acting, giving credence to the saying: "You can act your way into a new way of thinking faster than you can think your way into a new way of acting." Learning, teaching, documenting, and following up on specific expectations for focus on process are the first steps in implementing Lean management and developing a Lean mindset.

Study Questions

1. If you have had consulting help with Lean, how effective has it been and why?
2. How are people trained in Lean? Has the training resulted in skills in use and specific improvements?
3. Some gemba walks focus on teaching and learning. Others, even though called gemba walks, are actually inspection tours, adding items to someone's to-do list. Which is true here? How effective are they in increasing knowledge and practice of Lean? In creating an atmosphere of improvements?
4. Is a Lean or Lean management assessment used here? Is it freely accessible by anyone interested in using it? Has it been used to drive improvement and learning? If not, why, and what would have to change?
5. Have efforts been made to gemba walk with executives? Have they succeeded? Why or why not?

Chapter 8

Being the Sensei: Engaging Your Executives in the Lean Initiative

The first two editions of this book described a Lean management system as the missing link needed for line production managers to sustain Lean production conversions. Lean management was largely aimed at leaders responsible for day-to-day or tactical operations in areas converted to Lean production. Lean management as the missing link in Lean has gained traction in the past ten years. It has helped many Lean conversions last by effectively addressing a widespread cause of failure to sustain even objectively successful Lean initiatives.

But, as with other steps on the Lean journey, solving one problem has revealed another, one that threatens even the most successful Lean conversions committed to the behaviors, practices, and tools of Lean management. The problem is faded, weak, or nonexistent executive engagement with the Lean initiative, a widespread source of frustration in the Lean practitioner community.

With a Lean management system in place, the gains from Lean production can last and grow. But the Lean initiative remains at risk without the active engagement, involvement, and support from senior executive management. Most internal Leansters know that a loss of senior executive support will starve the Lean initiative of resources, badly damage its internal legitimacy among current and potential clients, and leave the Leansters, and their careers, stranded. This concern has emerged especially in the five years since the second edition of this book was published. During that time I have heard more and more internal Lean leaders express frustration about the difficulty of getting or holding executives' interest in Lean initiatives despite significant improvements in safety, quality, delivery, and cost. I have observed this problem in organizations I have worked with as a consultant, and I experienced it personally as an internal Lean leader. It was in response to fading executive interest in a Lean initiative that my Lean team and I developed the approach described in this chapter. You might think the kinds of outcomes from technically successful Lean initiatives, like those described in Chapter 7, should speak for themselves. They don't, at least not without involving executives in a way that is meaningful to them.

The original Lean management system, the tools, practices, and behaviors for line or tactical managers, did and still does represent a missing link in Lean. But, a second missing link has since emerged: an effective, repeatable approach for engaging executives with Lean *on their own terms* in ways that connect with their unique strategic and managerial responsibilities. It is likely that the threat represented in this second, executive missing link was always lurking, waiting to bring down the next organizational change program, whatever it was. For Lean, the engagement problem among those responsible for strategy has come into view more clearly as Lean management has sustained the gains in Lean initiatives. That is, Lean success at the tactical level has exposed the obstacle of failure at the strategic level to engage senior leadership. Sustaining the gains from Lean production has proved to be insufficient to sustain corporate-level support for the Lean initiative. Building executive engagement is the next obstacle, the next hurdle to overcome for sustaining and extending the gains from technically successful application of Lean tools and the Lean management system.

This chapter covers:

■ Symptoms of weak or fading executive engagement
■ Flaws in gemba walking Lean projects with executives
■ A framework for understanding the problem
■ The nuts and bolts of an approach to engagement proven effective and personally meaningful to executives
■ The tools, structure, and process for gemba walks that engage executives

Symptoms: Orphans, New Sheriffs, and the Next Big Thing

It is not unusual for Lean to get its start in the middle of an organization as noted in Chapter 7. Lean eventually catches the notice of executives, but often remains an orphan program, on the periphery of the company's strategic programs. In effect, Lean advocates may succeed in getting *permission* to continue pursuing Lean, but not true full-fledged support. Here, permission may carry the unstated condition of support contingent on Lean not interfering with other, formally sponsored initiatives. This might actually mean *no* support or, at best, weak support. It certainly is not the commitment for strategic initiatives that originates at the top of the organization, from the executives themselves (Case Study 8.1).

CASE STUDY 8.1: WHY GEMBA WALK THE FRONT LINE

The executive VP for finance and IT of a large healthcare organization asked why he needed to gemba walk at the front line of his organization. "That's why I have managers, to oversee those activities." Of course this is true, but Lean was new in this organization. How well did the VP's subordinate managers understand Lean? Their exposure to Lean was a training course two of the VP's managers had attended. In the training, they were taught how to write an A3 for a project in their own area. The VP declined to gemba walk. Instead, an IT director and I gemba walked the project area. We talked to the two frontline analysts and looked at the visuals they had put together to count issues to be resolved.

The visuals were tally charts on small whiteboards. They showed the number of problems outlying units experienced trying to log in to the central IT system. The charts in the two workers' cubicles made it easy to see the number of problems day by day during the week. The workers said they thought the tally charts helped them understand where to focus. The immediate manager of the area and the director said they thought the charts were worthwhile, though neither could identify how the data were being used to drive improvement. In fact, the tally data never left the cubicles of the two workers who recorded the information.

The organization had invested a quarter of a million dollars in the training. From what I was able to see, there was little return in the form of process improvement from that investment.

Had the VP spent an hour on the floor, it would have been clear that reports he received on the Lean project reflected *activity*, largely the training and A3 projects, rather than *improvement*. It is true that he had managers to oversee Lean projects in his area. But they had had only brief exposure to Lean in a training class. Understandably, none of them had learned enough to coach the frontline projects, or to realize the purpose of the charts should have been to convert the tallies into cause analysis, task assignments, and improved processes beyond what they had always done.

In a large organization, Lean takes time—one to three years—before its effects show up on corporate or even unit financial statements. Executives want to drive change and improvement. How can they tell if the Lean strategy they endorsed is working, especially if they are not interested in the nuts and bolts of frontline Lean production tools? The management system provides an early, nontechnical window into the effectiveness of Lean operation. It allows executives to reach their own conclusions about Lean, and to take action when they see deficiencies in management system implementations.

Or, you may have experienced the arrival of a "new sheriff in town." This is a newly promoted or hired senior manager unfamiliar with Lean who wants to make his or her own mark with a different approach in operations.

Or, your senior leaders at first really did support Lean, but now are inclined toward some different approach. Fads and fashions come and go in the world of large organizations, just as they do elsewhere. There is always a next big thing. It might be lights-out robotics, or enterprise software that promises to schedule procurement and production by remote control, a business professor's new book, or another consultant promising spectacular results from yet another new system (Case Study 8.2).

However it shows itself, the problem is rarely about Lean's results. Instead, Lean's failure to excite the interest of senior executives is more about the lack of genuine, robust, challenging involvement of senior executives in the Lean initiative in a way that is personally meaningful to them. This means senior executives directly participating in supporting and extending the Lean initiative. This kind of robust engagement among execs reinforces the benefits realized from Lean and those yet to come. Executives can more clearly see Lean as an important element in the organization's competitive strategy. In such an environment, the perceived need for the next big thing (and a new sheriff) is greatly diminished.

CASE STUDY 8.2: ENGAGE EXECS BY ACTIVE INVOLVEMENT IN LEAN

The VP for global business improvement and his colleague, a director for operations, were frustrated by their inability to engage the executive leadership team with a Lean production and Lean management initiative that slowly but surely was transforming the oldest plant in the company. The plant had a long history of customer complaints, but that was changing as the director and the plant learned how to apply Lean principles to the production process.

The VP had persuaded the president, his boss, that Lean was the answer to maintaining an edge in their highly cost-competitive business supplying a major consumer products producer. With the president's support, the VP had been delivering training in Lean concepts to the president and his executive leadership team two hours monthly for a year. But a recent field trip to the plant by this executive team had been a bust.

The various VPs and directors, even those with a technical background, got nothing out of the trip. They were bored on what was termed a gemba walk but was more of a show-and-tell tour of recent Lean-related changes to the highly technical, capital-intensive automated production process. The changes, early in the Lean effort, were meaningful to the plant's production leadership but obscure to their visitors. Lean had not yet yielded improved operating costs. Although technically worthy, the Lean effort failed to excite the interest of the visiting leadership team. Shortly after this disappointing plant visit, the VP and operations director asked for advice on how to engage the corporate leadership team.

I visited the plant and saw the early incremental process improvements and the beginnings of visual production tracking charts in use. I also attended two of the VP's ongoing monthly Lean meetings with the executive team to get a sense of the group. In the second of these meetings, I proposed a return gemba walk, this time structured by gemba worksheets for visual controls. The structure of the visit would have different pairs of the leadership group tasked with finding the answers to a subset of the worksheet's diagnostic questions in different areas of the plant. (The first failed plant visit had been nearly a year earlier. In that time, the plant had made further progress with Lean and with visual controls.) The executive team agreed to visit the plant again, but this time on a mission to assess the effectiveness of the initial implementation of the Lean management system.

I had met briefly with the president after the meeting with his leadership team. He seemed eager to get involved with Lean in a more active way. It was in this context that he asked a memorable question: "How can we be better leaders?" He meant, I'm sure, how can we better lead this Lean initiative that we have taken on faith to be so important to the business? I suggested the structured gemba walk process might hold part of the answer.

The date was set for the gemba walk, to take place on the day of a quarterly executive business review due to be held at the plant. After the business review, we paired up the executive staff members, assigned them guides to make sure they got to the right areas of the plant, handed out the Lean Management Gemba Worksheet for visual controls with highlighted diagnostic questions, briefed them on their task, and set them loose.

It was remarkable to observe. The pairs of executives were intensely focused on finding the answers to their highlighted questions, to the point of amusing several of the floor workers and supervisors they talked with. They practically ran from one assigned area to the next. They took extensive notes. They had heated discussions with each other about what ratings to assign. They listened intently and questioned the rationale when their Lean coaches critiqued the ratings they gave to their assigned areas. It was a demonstration of involvement with a capital I and engagement with a capital E.

The Lean initiative continued with good effect at the old plant, and later spread to two newer plants that were added to the responsibility of the old plant's operations director. My client, the VP, indicated that the structured gemba walk had broken the ice for the executive staff group. They felt they had a tool and process to guide their assessment of what they saw and heard when visiting Lean project areas in the company's plants and offices.

When Gemba Walks Are Not Enough

I was leading an internal consulting team bringing Lean applications to end-to-end business process value streams. The initiative had been the idea of the senior VP for global operations. His area's performance results had dramatically improved in five years of concentrated Lean applications in manufacturing, sustained by adoption of Lean management. He persuaded the CEO and his senior executive peers that Lean on the enterprise business side of the company could deliver similar benefits. The CEO named a steering team for this effort: four of his eight direct report senior VPs, including the operations executive. I reported to one of the four.

As the initiative was getting started, we met monthly with the senior steering team. After six months, we traded the hour consumed by the review meetings for a monthly one-on-one gemba walk with each of the four. The gemba walks appeared on the executives' calendars monthly, scheduled in advance for the year.

I met regularly with my boss, one of the four steering team members. About a year into the gemba walks, at one of our regular meetings, she provided some valuable feedback. "David," she said, "you have a problem." She went on to tell me the steering team members were losing interest in the Lean enterprise initiative. It did not matter that the project teams we taught, coached, and supported were resolving long-standing problems the company had not been able to get its arms around, some for as long as 20 years. The numbers—the 50 percent reductions noted in Chapter 6— did not matter. The problem, she said, was the senior executives did not feel involved in the initiative. It wasn't just an issue with Lean. "We tend to lose interest in initiatives after about 18 months", she told me. "Then we go looking for something new we think will help," she said, referring to the dreaded next big thing. She concluded: "You need to find a way to involve them."

I thanked her for her frank advice (for which I remain grateful!) and took the message back to the team. Reviewing our records of the past year's executive gemba walks corroborated her concern. Exactly 50 percent of the year's gemba walks had been canceled by a steering team exec and never rescheduled. None of us were happy with the gemba walks, and clearly our execs weren't either. The walks were our sponsors' primary contact with the initiative, and they were voting with their feet, in a direction moving away from us and the initiative! We needed to understand the problem and do something about it, and soon.

You may recognize this situation. It is one many internal Lean leaders have encountered. Gemba walking is how most of us learned Lean. Without gemba walks, Lean leaders are frustrated. "I want to gemba walk with my executives to teach them Lean, but after a couple of times, they stop. They won't do it. 'Lean is your job,' they say."

As Lean thinkers, we need to ask why, identify the root cause for lack of engagement, and develop solutions.

A Framework for Understanding the Engagement Problem

I had not been comfortable with the way we had been gemba walking. Compared to my experience gemba walking in manufacturing, our steering team executives were not engaged by the experience we were providing. When a scheduled gemba walk came up, our team scrambled to think of a project area to show them. We had no plan, no agenda. We simply took them to the project area and turned the project team loose to present to them.

In fairness to the executives, we had no reason to expect them to be knowledgeable about Lean concepts, value stream maps, Lean tool applications, and definitely not the foreign terms we and the teams used. And, the execs knew little about the teams' work processes, which happened five or six organizational levels beneath them. So, unsurprisingly, they did not engage with the teams in discussing an unfamiliar area and unfamiliar technical Lean topics. Instead, they did what they were good at and comfortable doing. They listened patiently, congratulated the teams for their good work, or hard work, or for what they learned, or whatever, and talked about the business. They were doing what a socially skilled leader does in a public appearance when there is no real agenda. No wonder half the gemba walks were being canceled.

We knew we could do better.

We addressed a number of questions aimed at understanding why our gemba walks had failed to excite the interest of our executives. These key points emerged from reflecting on our gemba walk experiences.

It became clear to us for the first time that we were in a supplier-customer relationship with our executives, and that we were the supplier. Value is defined from the point of view of the customer. Obviously we had not been providing value. Our objectives for executive gemba walks were the same as when we were learning Lean: gemba walk the way our teachers did to teach Lean concepts and implementation of Lean production tools. But our approach was not working.

Seen in the perspective of a customer-supplier relationship, defining the problem was straightforward. Our executives were just not interested in learning the nuts and bolts of implementing Lean production. Besides, they had people like us (and perhaps like you) to worry about those things.

The concerns of senior executives, what keeps them up at night, include strategy and its effective deployment through the organization to sustain or produce improved results. This led us back to the basic premise of Lean management: Lean production needs Lean management to sustain it. If executives mastered the Lean *management* system, they would readily be able to assess the fidelity of execution of the Lean strategy, all the way to the floor level. The ability to make this assessment rests on the proposition:

Healthy Lean management system = Healthy Lean production system

We concluded learning to assess the health of the Lean management system would be a good fit with our execs, meaningful to them in their own terms and attuned to their concerns. A problem in Lean management, after all, was a symptom of a management problem, undermining a key strategy our executive steering team members had signed on to support. As shown in the composite example in the next section, Lean management problems are ultimately executive management problems to resolve.

We considered what we knew about our executive steering team members. All the Lean team members had worked in one or more of the execs' organizations. What were they like as people? We needed gemba walks that fit with their personal attributes. This is how we characterized them as a group (which is probably similar to your senior executives):

- Bright
- Quick study, fast learner
- Competitive
- Achievement oriented
- Got an A on every test they ever took
- Seek challenge
- Desire to drive change and improvement
- Action oriented; thirsty for hands-on
- Expect staff prep work; be prepared and avoid embarrassment

Redesigning Executive Gemba Walks

From this, we realized a more effective gemba walk design would tie in to our executive customers' high achievement orientation, competitiveness, action orientation, desire for challenge, and desire to drive change and improvement. It became clear to us that our executive customers must have found our previous gemba walks dull and boring! Clearly, we had an opportunity for improvement.

Experiential Learning

The best way to learn Lean is through experience. A new gemba design had to provide our execs with experiential learning. (They had already told us they would not attend Lean training sessions or conferences.) We wanted a design that would give the execs a task to complete on the gemba walk, tapping in to their achievement orientation and competitiveness, and our conclusion about their desire for some kind of hands-on involvement. We wanted to put our action-oriented executive customers in charge of the gemba walk, responsible for their own learning, and accountable at the end of the walk for completing the specific learning objectives.

Repeatable Process

We realized we should design gemba walks as a repeatable process, one in which the execs would feel comfortable and prepared, where they knew what to do and how to do it. This is a key point: we realized our executives had been lost on our prior gemba walks, not knowing what to do or how to do it. We wanted the execs to feel in charge and know how to take charge. We wanted them to be in an active mode, not a passive recipient of a team's presentation. And, we wanted to work with a stable, predictable, repeatable process so we could better focus on coaching our executive apprentices and they could better focus on the task at hand.

Lean Management as Structure

I had been struggling to find a way to integrate the Lean management standards (in Appendices A and B) into our executive gemba walks, and this seemed like a good opportunity to experiment. The standards held the promise of transforming the unsystematic approach we had stumbled with into a well-defined and structured process. We decided to start with the three core standards (visual controls, accountability processes, and leader standard work), one at a time, beginning with visual controls. The identically formatted one-page standards with their diagnostic questions could serve as our staff work, quickly preparing the executive for the gemba walk. The standards represented a process—agenda, content focus, task, and structure (finding the answers to the highlighted diagnostic questions and assigning a rating)—and a proficiency test (did you see what the sensei saw? how did your rating compare with the sensei's?).

Visual Controls / Cycle Tracking: Lean Management Standards gemba worksheet
Location _____   Shift _____   Date _____

Intent: Visual controls should do at least one of two things:
•   Reflect the actual vs. expected pace or progression ... etc.
•   Capture delays, interruptions, and frustrations ... etc.
Diagnostic questions:
1.   Can you see visual cycle or procedure tracking ... etc?
2.   Are the charts etc?
.
.
.
.
7.

See Appendix C for complete set of gemba worksheets

Assessment: Rate this area / areas from 1 to 5 on the scale below and note rationale ... etc.

1: Pre-Lean	2: Starting	3: Recognizable	4: Stabilizing	5: Sustainable
No visuals ... etc	Some cycle charts ... etc	Many front line ... etc.	Visuals used for ... etc	Visuals regularly ... etc
Rationale for this rating:				

No=0%, Few<25% Some <50% Many >50% Most > 75% All = 100%

Figure 8.1   Example (format only) of a gemba work sheet.

(Since developing this approach to executive gemba walks, I have reformatted the Lean management standards for ease of use as a teaching tool—Lean Management Gemba Worksheets, found in Appendix C and available at www.dmannlean.com.) Figure 8.1 shows the format of a worksheet in schematic form.

Nuts and Bolts of Executive Gemba Walks

Consider this composite example drawn from experiences in several organizations: on a gemba walk, an executive encounters frontline workers, a team leader, or a supervisor unable to explain, for example, the reason for a posted production tracking chart. Or the chart's entries for reasons for misses are general, vague, ambiguous fragments rather than clearly described problems. Or the daily chart is out of date, still posted but dated days or even weeks earlier. The executive, skilled at assessing the quality of Lean management practices, recognizes he or she has come upon a problem.

What kind of problem is it? To start with, it is a sign of ill health in the Lean management system and, as such, a signal of possible problems in the Lean production system. But it is not a Lean production problem.

Instead, it represents a broken or weak link in deployment of the Lean strategy. Somewhere in the organization above the location of this gemba walk, adherence to the Lean management system is being allowed to slide. The partial redundancy built in to leader standard work is intended to guarantee the integrity of the standardized Lean production process. That guarantee has broken down. It is only a matter of time before disciplined adherence to Lean production begins to slide and negatively affect results, if it has not already.

Strategic Breakdowns Are Management Problems

So, this is a management problem, a problem in disciplined execution of the Lean strategy, a failure located somewhere in the chain of command. That makes it a problem ultimately for executive management to diagnose and resolve. In the meantime, should the gemba walking executive reprimand the worker or team leader or supervisor (or manager, director, or VP when that is the case)? No, the executive who has mastered Lean management can and should explain what he or she expects to see in a production tracking chart, how the charts are a first step in process focus and continuous process improvement, and talk about how those improvements mean better service to its customers or patients, or how they strengthen the organization's competitive position, and given the production problem, what might make a better problem statement in the "reasons for misses" column on the production tracking chart.

Following the gemba walk, it is the executive's job to find and treat the weakness in his or her chain of command, perhaps with the help of the Lean resource. In fact, the Leanster may have chosen this area to gemba walk precisely because it has problems (see Case Study 8.3). Or, the executive might choose to gemba walk the managers in his/her chain of command, starting in this same area and working link by link up the chain. The objective is to diagnose the location of the weak or broken links. In other words, the executive is testing those in the chain to learn who does not recognize the problem with the tracking charts. Who sees what the exec sees in the tracking charts, and who does not? Who can explain the purpose of and technique in completing and acting on the visuals and who cannot? On these diagnostic gemba walks, the exec has the opportunity to reinforce what he or she expects to see and why. And, this area and others under the same leadership group will no doubt be scenes of repeated gemba walks to be sure weakness does not appear elsewhere, and that the message—and lessons—have been received.

CASE STUDY 8.3: DEVELOPING EXECUTIVES' EYES FOR WASTE

The order management department had been working at Lean improvements for two years, but the department director and managers resisted implementing visual controls. Finally, they created something formatted like an accountability board on which customer service reps posted problems. When they had an order that went on hold for any reason, the reps recorded the problem on a sticky note, put it next to their name on the date the problem occurred, and left it there until it was resolved. The boards were exposing problems—good! However, once the problem situation was resolved and the order released from hold, no action was being taken to understand the underlying causes of the problems or to follow up in any way. When we asked the manager in the area what happened with the accumulated sticky notes, she told us she was saving them and that "eventually they will be compiled." This went on for some time.

As we gemba walked in the area, we asked the order management director about problem solving. He gestured around the floor, saying, "Everyone here is a problem solver!" As we talked with him, we observed that the same kinds of problems occurred over and over. The service reps did a good job of minimizing the impact of problems on customers, but they left the root causes undisturbed.

We brought each of our senior executive Lean steering team members to gemba walk the order management visual boards. The executives were by now familiar with visual controls. On these walks, we handed the executives the Lean management standard for root cause problem solving. On the gemba walks, the executives acknowledged and reinforced the manager's and director's consistent use of the boards, but also asked what happened with the information from the sticky notes once the immediate problems were cleared up and the orders released from hold.

In a conversation with the VP responsible for the order management area shortly after this series of gemba walks, she told us Pareto charts of problems that had caused orders to go on hold were now being brought to her weekly accountability meeting and assigned for cause analysis and eventual root cause corrective action. The leaders in the order management area had not just gotten the message to do something, but they had actually been taught the next steps by the questions they had been asked by the gemba walking executives, all but one of whom was from a different sector of the business.

The results of experimenting with the new gemba walk design for executives were dramatic, even startling. When given tasks to complete within a clearly defined structure, our executives were all over it, getting into the gemba walks with great intensity. They had questions to answer and a rating to assign, and they were determined to get it right! The one-page standard was all the staff prep work they needed, using the diagnostic questions not as a script, but rather as a starting point for conversations with people in the area, and for detailed observations.

Importantly, my experience with executives in other companies has been consistent with the results of our initial internal experiment. In organizations where in I didn't have 20+ years' experience in the firm, I have found the same response in executives I had only just met and to whom I was an unknown outsider, as described in Case Study 8.2.

In the original experiment with the new gemba walk design, it was as though a switch had been thrown. Our executive sponsors, from the first gemba walk of the new design, acted as confident observers in gemba, guided by the worksheet. And, after just a few walks, they became competent in the core elements of the Lean management system—which after all is neither complex nor technically difficult. They asked why one or another Lean management application was not being used in their area of responsibility (when the walk took place elsewhere). After only a few walks on an aspect of Lean management, they were comfortable delivering spontaneous coaching on the floor when called for by the situation. And, they soon began gemba walking their subordinates using the same process they experienced on the revamped gemba walks with us (Case Study 8.3).

Two elements of the new design were particularly effective. One was the test, comparing the executive/apprentice's rating of the area with that of the Leanster/sensei. They were highly motivated to learn to see what their sensei saw, that is, to ace the test. In addition to enjoying the new structure and process, the test made the experience personally meaningful, connecting with their achievement motivation and competitive nature. Doing well on the test gave them confidence, evidence that they knew this material, could make it their own, and use it themselves.

The second element was the repeatable structure and process provided by the gemba worksheet. In the revamped version of a gemba walk, as we met our executive/apprentices for a walk, several of them would ask: "What [Lean management] standard are we looking at today?" The one-page standard—now the Lean Management Gemba Worksheet—provided the preparation, process, structure, content, agenda, and test for every gemba walk. And, the worksheets made it easy for the executives to gemba walk their subordinates as they mastered a particular element of Lean management.

The best evidence of the turnaround was the cancelation rate for gemba walks following the redesign. It went from 50 percent cancelations in the old process to 0 percent (zero!) in the new design for the three years in which executive gemba walks continued. Another piece of evidence is the transferability of the process to other organizations, as noted above.

"New" Executive Gemba Walks: Tools, Structure, Process

There are many "right" ways to implement Lean and Lean management. Principles are important; orthodoxy, in my opinion, is not. So, the following is not a recipe for the single best way for engaging executives by gemba walking for Lean management. My suggestion is think of it as a starting point for developing your approach to involving your executives with your Lean initiative, learn from your experience, and continue to refine your approach accordingly. In other words, plan, do, check, act. For example, see Case Study 8.4.

Orient your executive student with a summary overview of Lean management. You can ask him or her to read this book, but a better way—or at least a lower hurdle—would be to create a one-page handout (for example, in PowerPoint notes view) to review with your student on or before the first gemba walk. Include the Lean management icon (Figure 1.1) or something like it. Below it, in the notes section in easily readable type, briefly summarize Lean and why Lean management is important. Here are some items you might include on a handout:

1. The objective of these gemba walks is learning to assess the health of the Lean management system.
   a. Lean is first about finding waste, then eliminating it.
   b. Visual controls bring close focus on the process, in part to capture instances of waste—problems, interruptions, and delays—as they happen.
   c. An accountability process converts the problems caught on the visuals to assignments for action, first to understand root causes, and then to eliminate them. This improves the process.

   d. Leader standard work maintains the visuals and accountability
      processes, and monitors to make sure recently implemented
      improvements are understood and consistently executed.
   e. When the management system is healthy, you can be confident the
      production process is in good health and improving.
2. As you begin your first gemba walk, orient your student to the Lean
   Management Gemba Worksheet (Figure 8.1 and Appendix C).
3. Briefly orient the student to the worksheet.
   a. At the top, the name of the LMS element.
   b. Next, the intent of this LMS element.
   c. Then, the diagnostic questions (today's are highlighted) with space
      for notes.
   d. Followed by the rating scale (1–5) with characteristics of each level.
   e. Finally, space for noting the rationale for the assigned rating.
4. Hand the worksheet to your student: "Those are today's questions; let's
   find the answers!"
5. Coach the student to open with the purpose of the visit: "I'm here as a
   student to learn how you use Lean management. Jim/Jane is my coach.
   Do you mind if I ask some questions?"
6. Make introductions in the area as needed, then let the student lead.
7. As needed (but sparingly!), ask a leading question to support the
   conversation.
8. At the end, thank those in the area and ask, "Do you have questions
   for us?"
9. As you walk back, ask your student how he or she rated the area. If his
   or her rating differs from yours, you have about a 90-second teachable
   moment to point out what you saw and why you gave that rating.
   Be direct: you only have 90 seconds! When the student's rating matches
   yours, congratulate the student and briefly compare each other's rationale.

**CASE STUDY 8.4:   A BURNING PLATFORM
FOR LEADER STANDARD WORK**

The synthetic crude oil plant of an integrated petroleum producer had
experienced two catastrophic fires in a four-year period. The incidents
shut down half or more of the plant's—and thus the company's—output
for months at a time, and depressed its stock price for years. The company had experienced dramatic growth, expanding production of the raw

material that supplied the plant. But as the plant's output was increasing, many in the workforce who had grown up with the plant reached retirement age, or left for other reasons, including the site's remote location and severe climate.

With the workforce turnover, much of the informal or "tribal knowledge" accumulated by the veteran operators left with them. The fires, among many less severe incidents, were the result. Inexperienced workers and supervisors with only cursory training were left to operate a major, complex chemical plant with minimal knowledge and little background upon which to draw. The company had been following a recipe for disaster. Unfortunately, it had the disasters to show for it.

The company's response was vigorous. It included outside hires of two key executives. One was a senior VP who had been site manager for a major petrochemical complex and also led that site's successful Lean conversion. The second new executive had managed a number of chemical and petrochemical plants in his career and, importantly, was an expert in the process safety management protocol used in process industries worldwide. The senior VP asked me to support implementation of Lean management as part of the recently begun Lean conversion of the site.

I gemba walked in the plant's operating areas to understand how the process was managed. I learned what "focusing on the process" meant in this multiacre, potentially toxic and explosive operation. Shift supervisors sat in a blast-proof sealed control room watching computer displays of process operating readings. They rarely ventured into the plant itself. I also learned the detailed operating procedures for the plant were often obsolete, rarely consulted, and even more rarely observed. It was pretty clear that operators and supervisors were, in effect, living by their wits and doing the best they could.

Part of the Lean transformation involved developing standardized work starting with standardized routes for operators to take through the plant to make the observations and perform the few manual tasks included in a normal shift. A parallel task was development of standardized work and routes for the supervisors. A third element was development of a one-page checklist of key details for monitoring adherence in six critical process and procedure requirements. The checklist called for in-person observations and questions for operators at multiple stops throughout the plant. These included the area that, by its nature, carried the greatest risk of uncontrolled release of toxic fumes and explosive vapor. Operators in that area had told me, "The supervisor doesn't come here. Ever."

The one-page checklist was the equivalent of a Lean Management Gemba Worksheet with questions to test operators' and supervisors' understanding of procedures and standards, and observations to check for adherence with them. It was developed for two reasons. One, it would clarify expectations for supervisors and operators. Two, an explicit purpose of the checklist was to prepare managers and executives in the chain of command of the plant and the site to be competent and confident assessors of the manner in which the plant was operated. From the executive VPs on down the chain, managers were expected, for the first time, to be able to personally verify from their own knowledge gained on the gemba walks that the plant operators, supervisors, and immediate line leaders understood critical operating procedures, knew why they were important, and went about operating the plant safely, reliably, and efficiently.

The site senior executive leadership team understood the intent of leader standard work, to guarantee the integrity of the standardized Lean or, in this case, the safe and reliable operation of the process. Supervisors understood the checklist-defined expectations for good practice. A new expectation for operations managers (to whom supervisors reported) up through site senior VPs was to gemba walk in the plant regularly, guided by the checklist to focus their observations and conversations. There was no test to ace in this version of a gemba worksheet, but there definitely was a burning platform, literally and figuratively, for maintaining close focus on critical processes.

The company had turned over a new leaf in its attention to safe and reliable operations, taking many steps in addition to the ones described here. Most importantly, they understood that leaders', including executives', responsibilities included demonstrating and reinforcing close focus on and adherence to defined processes by personally following standard Lean advice: go to the place, look at the process, talk with the people.

Summary

The Lean management system as described in the first two editions of this book has been adopted by many organizations to sustain their Lean production initiatives. It has indeed proved to be a missing link in Lean. But even with sustained Lean successes, many Lean practitioners find their executives have lost interest in and support for the Lean initiative. Perhaps surprisingly, this often includes execs responsible for operations. Lean management's positive impact on sustaining Lean does not resolve this paradox: despite its continued success, executives in many organizations lose interest in Lean and turn to some other change or improvement strategy. Lean and its practitioners end up isolated, their initiative loses resources, and Lean eventually fades just like other "flavor of the month" programs.

Many internal Lean resources attempt to build executives' support by gemba walking Lean project areas. Their reasoning seems sound: "Gemba walking with a teacher, coach, or sensei is the way I learned Lean. That's how I'll develop Lean thinking among my executives." Unfortunately, in most cases this approach does not work. First, most senior executives, even with technical or operations backgrounds, do not want to learn to become Lean implementers. They have people like us to deal with the details of a Lean implementation.

If the Student Hasn't Learned, the Teacher Hasn't Taught

Executive managers, even those with training or experience in operations, are, for the most part, no longer focused on the day to day, on the tactical. Instead, their responsibilities include longer-range objectives, including developing the organization's strategy and the integrity of its deployment. Engaging them in Lean at this point in their careers will take something different from what engaged most of us who are Lean proponents and advocates. Remember the Lean adage: "If the student hasn't learned, the teacher hasn't taught." It is up to us to find ways to excite executives' interest in Lean through their own experiences, observations, and benefits as viewed from their own perspectives.

Second, many executives in corporate functions outside operations are unfamiliar with and somewhat uncomfortable on a production floor, be it manufacturing, administrative, or service delivery. They may have worked in that kind of setting earlier in their careers, but technologies, techniques, tools, and products have likely changed since their days on the floor.

Third, most executives are not fluent in Lean. The jargon is probably foreign to them. And, they lack the knowledge and practice to assess for themselves the quality of a Lean application. They may have been enthusiastic about Lean initially, but they do not know what or how to actually support Lean as part of their normal responsibilities. They do not experience Lean in a way that involves them and excites their interest, so they turn to the next sure-fire strategic idea, the next big thing they believe will help them drive change and improvement.

Fourth, it is common for the interest executives have in most organizational improvement initiatives to fade after 18 months. The way to halt and reverse this phenomenon is to find a way to involve executives in the initiative in a way that contributes both to the initiative and, given the nature of their positions, to the organizational interests of the executives. Involvement based on experiential learning is more engaging by far than exposure, no matter the quality, to concepts, success stories, and simulations. These kinds of exposures lack the impact, and staying power, of personally meaningful learning experiences, especially when the experiences help executives achieve the goals that accompany their responsibilities.

Consider Executives as Customers

What is so frustrating for internal Leansters is that this loss of executive support and interest ignores Lean's often impressive improvements in safety, customer satisfaction, quality, delivery, and cost. What is the alternative to coaching the executives the same way you would teach Lean to anyone else? The answer lies in treating executives as customers, understanding what they value, their learning styles and other attributes, and the unique nature of their strategic responsibilities. On this basis, it is possible to design an approach that engages them, prepares them to be involved, and equips them to deal more effectively with important management problems that they alone can solve.

The key to this approach is that the long-term health of Lean production depends on a healthy Lean management system. Lean management is, after all, about management tools, behaviors, and practices. Consider these scenarios: Lean management is weak, incomplete, out of date, or poorly implemented. Or, frontline workers and leaders cannot explain its value. These are clear signs of strategy execution problems somewhere in the chain of command, problems uniquely for executive management to resolve.

Providing executives with a structured, repeatable process for gemba walking, first to learn, and then to assess the health of Lean management, equips them to be confident, comfortable, and actively involved in managing the execution of the Lean strategy they support.

Study Questions

1. Have you tried gemba walking executives to build or reinforce their support for Lean? Has it been effective? Why or why not?
2. Where on the organization chart would you draw the boundary between tactical and strategic responsibilities?
3. Does your organization have a history of trying one program after another to drive improvement? If yes, why?
4. Who are the key executives in your organization whose support you need for sustaining your Lean initiative? How would you describe them as people?
5. If you have conducted gemba walks with executives, who has taken charge on the floor: you, the exec, or someone else? Have the outcomes been satisfactory? Why?
6. Have executives refused to gemba walk?
7. What, if anything, surprised you in this chapter?

Chapter 9

Leading a Lean Operation

Leadership is a topic that applies more broadly than Lean management, to be sure. It is regularly discussed whenever the subject is organizational change. A Lean conversion initiative is properly and usefully thought of as this kind of change, as well as a change in the technical ins and outs of production systems. That is because good engineering does not by itself make for effective change, not by a long shot. Without effective leadership, most large-scale changes in systems do not go well and do not perform up to advertised expectations.

This is emphatically true for Lean conversions because, as I have suggested, a Lean conversion requires profound changes in the way you think, and in deeply established habits. Without determined leaders and effective leadership, no conversion project is likely to live up to what was hoped for it. For that reason, this chapter touches on nine dimensions of leadership that are needed both for leading an effective Lean conversion project and, in subtly but importantly different ways, for leading an ongoing Lean operation.

Nine Leadership Behaviors to Learn

Successful leaders are those who behave in particular ways. In other words, success is based on what you do, not on who you are. That is fortunate, because for most of us, it is too late to be a born leader! Behaviors, on the other hand, can be learned and unlearned. Included here are how you respond to interruptions in production, the way you arrive at conclusions, and what you ask people to pay attention to. Table 9.1 lists and briefly

Table 9.1   Dimensions of Lean Leadership

Attribute	For Project Implementation	For Ongoing Operations
Passion for Lean	1. Passionate about the potential for Lean to make the enterprise more successful and work more fulfilling for all involved.	1. Same as project implementation, plus: 2. Willing to make personal changes in one's own work, including using standardized work for his or her own position.
Disciplined adherence to process—accountability	1. Sets expectations, regularly uses a process to track and follow up on actual accomplishment of assigned tasks.	1. Same as project implementation, plus: 2. Exhibits intense commitment to focus on explicitly defining processes and disciplined adherence to them.
Project management orientation	1. Prior experience in successfully implemented projects. 2. Uses a defined process to track performance and completion of task assignments. 3. Identifies corrective action where necessary and follows up on it.	1. Able to identify needed changes based on daily process data and assign small-bite daily tasks leading to successful implementation of the changes. 2. Uses explicitly defined visual processes to track and follow up on assignments and take appropriate corrective action.
Lean thinking	1. Understands Lean concepts. 2. Has had experience applying Lean concepts. 3. Talks about and promotes a Lean future state. 4. Finds ways to apply and illustrate Lean concepts in daily project work processes.	1. Serious about ongoing improvement based on a goal of perfection. 2. Sees with kaizen eyes. 3. Holds and coaches a root cause orientation to corrective action. 4. Has learned process improvement/problem-solving methods; able to personally lead and coach Lean process improvement.

Table 9.1 (*Continued*)   Dimensions of Lean Leadership

Attribute	*For Project Implementation*	*For Ongoing Operations*
Ownership	1. Thinks and talks about the area as his or hers to lead, set direction for, change, and improve.	1. Same as project implementation, plus: 2. Eager to empower others in the area through structured ways to elicit and implement their ideas. 3. Acknowledges and celebrates improvements made by others at all levels.
Tension between applied and technical	1. Understands the need to sweat the details, as well as to get things done. 2. Willing to listen to technical experts and consider their advice in planning for the implementation.	1. Understands and respects the details behind elements of Lean, such as flow, pull, standardized work, etc. 2. Actively supports steps to upgrade performance and expose previously hidden impediments. 3. Takes a "What can we do today?" orientation to making change happen steadily, step-by-step.
Balanced commitment to production and management systems	1. History of effective give-and-take with people and ideas. 2. Evidence of process focus beyond a "hit the numbers" approach to management. 3. Eager for greater participation by production people as well as others.	1. Personally treats process focus as crucial to the area's success; is able to see waste and opportunity even in leaner processes. 2. Insists on compliance with requirements for visually tracking process performance and execution. 3. Insists on analysis and appropriate, timely action on impediments to normal operation of processes.

Continued

Table 9.1 (*Continued*)   Dimensions of Lean Leadership

Attribute	For Project Implementation	For Ongoing Operations
Effective relations with support groups	1. History of getting things done with support from operations support groups such as engineering, quality, production control, safety, finance, HR.	1. Understands roles, responsibilities, and expertise of support groups. 2. Incorporates support groups appropriately in plans for improvement and responses to problems. 3. Makes expectations explicit for support group performance in support of production processes.
Measure process separately from results	1. Breaks the project into small steps with due dates. 2. Frequently assesses and verifies completion status. 3. Acts to resolve problems right away as they come up.	1. Creates measures to frequently document process performance and misses. 2. Establishes regular, frequent review of process misses and trends over time. 3. Teaches and emphasizes cause analysis, root cause solutions, and connections with improved performance.

describes nine dimensions of leadership that we have found important in making Lean conversion projects and ongoing Lean operations successful.

Leading a Lean conversion project differs in significant aspects from leading an ongoing Lean operation, one that has already been converted or that started as a greenfield. The differences between them are discussed in turn, contrasting what is needed to successfully lead a conversion project with what is needed to successfully lead an ongoing Lean operation.

Attribute 1: Passion for Lean

It is probably true that passion for Lean is something more than mere behaviors that you simply learn and repeat. It can manifest in many ways, and importantly, in which charisma or force of personality is *not* required. Think of the people some might call geeks. A geek might be one seriously committed to a pursuit, who talks about it often in terms of possibilities that can make the future look better than the present, and draws connections about his or her subject and the external circumstances facing him or her or the group. That is someone, you would say, who has a passion, whether for long-distance running, religion, politics, or even Lean production. Right?

In passion for Lean, an important facet involves comparing Lean with the alternative. For example, Toyota in 2004 was reported as being seriously concerned about what it describes as a looming challenge from China, now being realized. It used that as motivation to improve further and further; this in a company whose market capitalization exceeded that of all its North American competitors combined. Fifty years ago, Toyota was motivated by the threat it perceived from American manufacturers taking over its domestic market.

It is easier to be passionate and enlist others when you have a case for change based on real external factors that are affecting or likely to affect your enterprise. There is something about a fight for survival or for your place in the market that is more likely to arouse passion than merely a drive for increased value for the shareholders or bigger bonuses for the executives.

Enthusiasm, intensity about a subject, willingness to engage others on their terms with respect to the threats and possibilities, deep knowledge about the subject, examples from one's own experience—all of these are marks of passion. These are attributes that can be studied, learned, and acquired over time. They grow from believing that there must be a better way for your organization to survive and prosper in a competitive world. I say this from my own personal experience with Lean. I'll bet it has been your experience with Lean or other things that have brought you to this point. That is why I contend that Lean leaders can be made. I doubt anybody, even Taiichi Ohno, was born "Lean." Ohno had to start his journey by studying Henry Ford!

Willingness to Make Personal Change

The best leaders of Lean operations I have known are also the ones who have recognized that the methods they insist on others using are also ones they themselves adopt. A principal difference between just leading projects, compared with leading Lean operations, is that leaders of ongoing operations have more, and more meaningful, opportunities to take the step of personally adapting themselves to a leaner mode of operating.

I have known advocates of Lean who have labeled and taped off spots on their desks for stapler, phone, and coffee cup, and some who label their inbox tray "WIP" and put a Post-it® on a page near the end of their current pad of paper to signal the reorder point. But perhaps the clearest and most persuasive example of this involves standardized work for leaders (see Chapter 3). Part of this is simply leading by example. Operators will often complain about having to follow standardized work when it is first introduced. When a team leader, supervisor, value stream manager, or plant manager can pull out his or her standardized work for that day and compare notes, it makes a powerful statement. "I use this so I don't forget to cover each of the departmental boards each day, just like standard work in a workstation can prevent assemblers from missing a step that might send a flawed product to one of our customers." That is a lot more persuasive than a statement that in effect says, "Because I have the authority, I'm telling you to follow standardized work." Beyond that, of course, is that a value stream manager who follows his or her standardized work will be far less likely to miss covering each of the departmental boards every day! The same principle—standardized, explicit expectations that help actual performance meet what was expected—is at work in both cases. The passionately committed leader recognizes that and makes the personal change to adopt new, leaner ways.

Attribute 2: Disciplined Adherence to Process—Accountability

Setting clear expectations and using a regular process to track completion is crucial for those leading projects to design and implement Lean production. Establishing accountability to task assignments and schedules is one of the basics in effective project management. The analogous skill in managing an ongoing Lean operation also involves establishing accountability, but in a much more dynamic, live production environment. Planning and defining tasks is still important, but must happen more quickly. These steps are guided by an overriding principle: integrity of processes. A project team has a finite, definable task and at least some time for investigation and analysis. In an ongoing Lean operation, the Lean leader's focus on his or her processes is what drives much of the accountability.

This process focus is a daily vigilance that begins with what can appear to be an obsession to explicitly define every last process in the leader's area of responsibility. These process definitions typically start with an applicable form of standardized work, for a production line, for example. Hourly or more frequently tracking and recording production versus goal tells whether the standard is being met, and if not, what appeared to interfere. That is a process as well. So is a third process, by which follow-up on reasons for missed takt or pitch cycles is regularly assigned in the daily accountability process three-tier meetings. And so is the process to check for completion of assignments on their due dates. That is four levels of definition for a single process. And this does not consider leader standard work as a fifth process to guarantee each of the four steps just listed! See Figure 9.1 for a graphic depiction of this example of a set of reinforcing processes.

In Lean management, virtually all processes have initial definitions and documentation. In addition, for each there are secondary processes to verify that the initial process was executed, and often further follow-up or verification processes. The periodic gemba walk of the plant manager by the operations VP is an example of the secondary verification process. The standard work of the leaders, from value stream manager down through team leader, and the related procedures extend the verification. The idea is

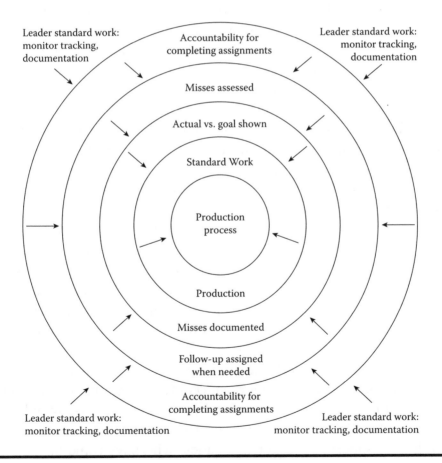

Figure 9.1   Example of multiple layers reinforcing focus on process.

to guarantee, as far as is humanly possible, the integrity of the production process, not simply to set it up and hope it all works as planned.

In many ways, adherence to processes is like the discipline of redundant quality checks in successive workstations in an assembly line. Standardized work in most stations begins with checking a critical aspect of the work elements performed in the previous station. It is the leader's role to establish this climate of discipline in definition and follow-up, in which the norm is accountability for faithfully executing the processes.

Attribute 3: Project Management Orientation

The necessity for effective project management skills when leading an implementation project hardly needs much comment. Regularly using a well-defined process to follow up for accountability on task assignments is important in a project environment as well as in ongoing operations. The ability to think in terms of work breakdown structures is another project management practice that translates into a daily operating environment. A work breakdown structure is nothing more than the step-by-step sequence of subtasks that, when assigned and completed in sequence, result in completion of a single larger task. Work breakdown analysis is just another way of talking about detailed step-by-step planning.

What is different in an ongoing operation is that the analysis and planning often happen in a stand-up meeting on the floor as the leader is scrutinizing yesterday's performance tracking charts and other measures of process effectiveness or interruption. The leader, as he or she examines the reasons for missed takt or pitch cycles, needs to be able to perform analysis quickly, which might be as simple as using the five whys (a basic method of cause analysis; see the glossary for a more detailed definition of the five whys). Next, he or she must be able to identify the initial steps to take, to further understand the cause, put a countermeasure in place, or initiate corrective action. Then, the leader needs to make an assignment generally of not more than a day's duration for a staff member or support group representative to take the appropriate action.

This is very much like performing a work breakdown analysis except that it happens quickly, almost but not quite on the fly, and often with people who have not been exposed to this kind of thinking because they have not been involved in managing projects. That is especially likely to be the case with floor supervisors and team leaders who have come up through the ranks. Over time, a leader using this process to address everyday interruptions, anomalies, and problems in the production environment can teach this step-by-step planful approach to his or her direct reports. For example, if a value stream manager (or supervisor) assigned a department supervisor (or team leader) the task of updating standardized work and rebalancing a flow cell when both the tasks and the process were new to the supervisor, the odds of success would be low. Consider this alternative:

- First, the value stream manager introduces the task, then he or she breaks it into constituent parts, such as updating standard work and updating the operator balance chart one position at a time.
- The value stream manager makes the tasks into daily assignments and reviews each one daily as part of the third-tier accountability meeting.
- Next, the leader does the same with the tasks needed to rebalance the cell.
- By the time the work is done, the supervisor has valuable hands-on experience with a basic tool in designing flow processes and has learned, or at least been exposed to, a work breakdown structure to get a complex piece of work done one manageable step at a time.

These project management tools represent valuable skills for an organization to impart to those in junior leadership positions. Value stream managers who have more formal experience in leading and managing projects are in a much better position to do this kind of teaching by example than those who do not.

Managing daily task assignments using visual controls is discussed in detail in the daily accountability three-tier meeting structure and process in Chapter 5. To summarize that process here, the meeting leader writes the task assignment on a Post-it note. The leader places the note on a names-by-dates matrix in the square that corresponds to the due date for the assigned individual to report on task completion. The names, depending on the level of the meeting, are those of supervisors and support group representatives (tier three), or of team leaders (tier two). Assignments are made, depending on the level of the meeting, by the value stream manager or by the department supervisor. Completed tasks are color-coded green, and overdue tasks are coded red.

The matrix is part of a value stream or department information center, a visual display and stand-up meeting place located on the floor in the production area. This visual assignment and tracking method highlights accountability for completing daily assignments.

Attribute 4: Lean Thinking

Lean thinking in an implementation project is like being a practical dreamer. A project leader needs to balance what we might do in an ideal future state with what we can actually accomplish given the scope, schedule, and budget for the project. Lean thinking in an ongoing operation takes a different tack based on an understanding that Lean is fundamentally an improvement system. In this respect, Lean thinking in the leader of an ongoing Lean operation is derived from the goal of attaining perfection in the operating system. This is something of a curse, a sort of "princess and the pea" situation, where, in the leader's eyes, things are never good enough and improvement is always possible. The leader needs to see where improvement can be made even if he or she cannot say exactly what the improvement should be. That, after all, usually requires the involvement of those with the deepest knowledge of the process in question, those who work in or with it every day. Nevertheless, if the leader has "kaizen eyes," always seeing something that might be improved, the rest of his or her organization will develop the same habits of perception, if only to keep a step ahead of the boss! (Case Study 9.1).

CASE STUDY 9.1: TAKING LEAN TO THE NEXT LEVEL

A case illustrating this involves the supervisor described in Case Study 4.4. His fabrication area had come to run smoothly, in part because of the use of job-by-job tracking. Through its use by the operators in the area, the most frequently occurring sources of interruption had been eliminated. However, the supervisor was not satisfied. He assigned the fabrication team leader the task of Pareto-charting the sources of interruption that remained after the initial cleanup had taken place. This was painstaking work, but there were clearly a few things that had moved into the position of most troublesome, and the leader was determined to find and eliminate them. The next level typically only becomes visible once you eliminate the problem that previously obscured it from view. The next level is usually represented by the next problem—the next opportunity for improvement.

Lean Thinking Looks for the Sources of Problems

A Lean thinker is interested in and in a way even welcomes (if not joyfully) the appearance of interruptions, anomalies, and problems in his or her processes as opportunities to understand and eliminate sources of variation and disruption. This root cause orientation to corrective action is a powerful engine in driving continuous, lasting improvement. After all, continuous improvement focused on the same set of recurring problems is not exactly what Lean thinking is all about. Instead, a Lean thinker's version of corrective action is aimed at uprooting the sources of problems so they never appear again. A Lean leader's capabilities should include being able to lead process improvement and problem-solving activities as an example to the organization and as a teacher and coach for these important tools and ways of thinking.

Attribute 5: Ownership

Ownership as an attribute of a Lean leader actually has more to do with enabling and celebrating the contributions of others to progress in the area. That is, a Lean "owner" definitely is responsible for setting and reinforcing direction in the area, just as the leader of a Lean conversion does. The Lean leader of an ongoing operation sets direction at a high level. He or she then creates conditions and specific structures and processes through which those in the area can participate in making changes that bring the area closer to the overall vision and direction.

In this regard, ownership is broadly analogous to presiding over a democratic process. The principles of Lean in this case will be scrupulously observed. The specifics of which changes are made will depend in considerable part on the ideas that come from the staff and from the floor. In other words, the owner does not dictate what changes will be made; rather, he or she acts to teach and challenge others to develop their own suggestions for how best to move in the established direction.

The examples in Case Study 9.2 are the kinds of simple but creative things you can expect when the owner of an area sets the conditions for those who work in it to feel free to make changes for the better.

Attribute 6: Tension between Application and Technical Details

A value stream leader from a traditional production background might be tempted to leave the technical details to the "experts," the engineers or local Lean resources. That is, a traditional leader might not take the time to learn

CASE STUDY 9.2: TAKING CONTROL OF IMPROVEMENTS

There will be plenty of instances where the leader/owner will have to assert the right to interpret what it means to be consistent with the principles. For example, one value stream sent separate schedules to work centers that happened to be lined up in the order of the production process. The value stream manager called this a pull system. He labeled as a supermarket the areas where the WIP and overstock accumulated between work centers. He referred to the expediters who sorted and picked through the stock as water spiders.

This is a clear instance of where a Lean owner (in this case, the plant manager), prompted by his Lean sensei, asserted his owner's prerogative. He helped the value stream leader better understand the application of Lean principles and the difference between a push system (no matter how sophisticated the scheduling algorithm) and a type B sequenced pull kanban system.

By contrast, you want to see examples where people have taken the teaching and general direction of the owner—for example, "Flow where you can, pull where you can't, and never push!"—and created something entirely new that solves a problem or improves a process in a way nobody, certainly not the Lean owner, had thought of. When the owner has established the right conditions, there will be many of these instances.

Two further cases illustrate the creativity possible when ownership involves the opportunity to share responsibility for improvement.

One involved a series of small welding cells in a metal fabrication area. It was difficult to tell the status of the cells by looking because nearly opaque weld curtains shielded them. The team leader used dowels to make simple flag poles extending out at a 45° angle from the top of the rods holding the curtains. If the green flag is at the top of the pole, the cell is running; red at the top means trouble; red and green even with each other in the middle means the cell is idle.

In the second case, the Total Productive Maintenance (TPM) facilitator in a machine area devised a simple method for documenting the daily tasks at each machine and at the same time visually indicating the status—completed or not. He took black-and-white and color photos of each task, such as checking a transport belt's tension. He annotated the photos to show the particulars of each task and then glued the black-and-white photos to the side of the machines' control panels. He put the color photos in magnet-backed, rigid plastic picture holders such as you might find on a refrigerator door at home, holding pictures of family or friends. He put the color photos on top of the black-and-whites under a header label that read: "Daily Tasks." When the operator completed the task, he or she moved the color photo off the black-and-white, placing it under a column header label that read "Complete." At the end of the day, the team leader moved the color photos back onto the black-and-whites and checked off a log sheet indicating the task had been done.

and understand machine balance charts and the implications for shrinking machine changeover times to reduce inventory quantities in a supermarket. Failure to appreciate the tension between the applied and technical poles in a situation like this can lead to commands simply to "cut inventory by 25 percent." In a system on the Lean journey, this is a surefire recipe for widespread stock-outs, one I have seen throw a system into weeks of chaos.

These are technical details that Lean leaders need to respect and be familiar with. It is true that there is a certain "just do it" mentality in the Lean drive to learn from action. It is also true that process integrity and performance can be compromised unnecessarily by a leader's failure to understand how Lean principles translate into well-defined capacities, replenishment times, and the like. It is true: Lean is not rocket science. It is also true that you cannot cheat physics: cutting inventory without reducing changeovers can easily result in stock-outs. Taking people out of a line carefully balanced to takt time and asking the rest to "work faster" will likely result in falling short of the day's production requirements, potentially incurring unplanned overtime, and putting safety and quality at risk. If that isn't bad enough, there is high likelihood of causing resentment among the frontline workforce and poisoning the reputation of Lean.

It is one thing when, because of specific conditions, a leader deliberately decides to take these kinds of actions with full realization of the likely consequences. It is an entirely different thing when the outcomes surprise a leader who has not learned how Lean works from the inside out.

At the same time, Lean leaders are most certainly in the business of taking risks. That is the essence of the drive for improvement. Lean improvements usually involve reducing some form of non-value-adding activity or resource deployment, for example, reducing times for setup and changeover, reducing floor space, inventory, or queue, rebalancing a line to run at takt instead of a fraction of takt, moving people out of a process, linking previously separate processes, or other measures. When they do these things, the leaders deliberately stress the Lean system and expose its weaknesses. Armed with this information, they then can set out to shore up the weak areas and stabilize the system at a new, higher level of performance. A leader cannot know what will go awry when these cushions are reduced or removed. For that reason, only one of these actions should be taken at a time, in part to isolate the cause of any resulting disruption, and in part to mitigate the risk to the system's ability to deliver for its customers.

Finally, Lean leaders look at every day as an opportunity to move toward better performance. A simple but powerful manifestation of this is the question "What can we do today?" when considering one or another approach to solving a problem or making an improvement. Analysis and understanding the mechanics of Lean are important to be sure, but the Lean leader sets the tone for action, even if for small steps. In this mode of leadership, it is often better to act rather than taking another day of analysis and inaction. Weigh the realistic risks (as opposed to doomsday fears), of course, but think of the step as an experiment, an actual rather than theoretical analysis, testing the proposition being considered. Lean is in many ways an everyday version of experiential learning, sometimes confirming what you anticipated, sometimes surprising you, and sometimes doing both simultaneously. Experience is the only way to learn what Lean really is; your actions provide the classroom in which the lessons are given.

Attribute 7: Balance between Production and Management Systems

Leaders in a Lean conversion project typically and appropriately focus most of their attention on the design details of the physical production system, and rightly so. It is up to the leader of an ongoing Lean operation to establish the Lean management system that sustains and extends the gains from the technical Lean implementation. Of course, if the Lean project leader also happens to be the leader who will run the area once the implementation is complete (a practice I definitely recommend), he or she will benefit from thinking about elements of the management system during the project phase.

In too many cases, the technical details continue to rivet the focus of the leader even after implementation of the technical design. This is a frequent cause of failure of the implementation to deliver on its promises, a cause far more common than technical miscalculation. The leader must not consider the job done when implementation of the technical design is complete. Instead, an effective Lean operations leader continues to look for, and invariably see, waste, and thus opportunity, in even the newly implemented processes. When the leader engages in constructive critique of his or her own work and focuses on opportunities for improvement, he or she sets a tone for the orientation to continuous improvement to take root and spread.

As important, the Lean operations leader brings an intense focus on process, specifically treating process focus as a crucial element in the Lean area's success. This focus includes personally following up on requirements for maintaining visual process controls for performance and execution:

■ Are hour-by-hour or other forms of production tracking charts being completed in a timely manner or in batch mode at the end of the day?
■ Can team leaders and supervisors speak knowledgeably about the flow interrupters or other process disruptions recorded on the charts?
■ Are the reasons for misses clear and specific enough to act as a basis for moving to the next steps?
■ Are controls in place for noncyclical processes, such as weekly 5S audits, operator-based maintenance tasks, periodic scheduled maintenance activities?
■ Are visual controls in place to track performance between processes, particularly the "re" processes: rework, reorder, refinish, and repair?

The visual controls and focus on process and reasons for abnormal operation will yield a trove of data on how a Lean area is working and the instances when it fails. Maintaining the visuals and tracking tools is work for many people. When they see their work taken seriously, used as the basis for analysis and problem solving, and when problems get fixed and interruptions are eliminated, it makes all the effort of the new routine worthwhile. So, the data must be analyzed, ideally involving to whatever degree feasible the people who generated and recorded the data. The analysis needs to entail using the smallest tool for the job (many problems do not require a designed experiment, for example, where a carefully conducted five whys will do). Most important, the analysis needs to result in action. Indeed, the maxim from Kurt Lewin—"No action without data; no data without action!"—applies here.

This is a lot for a Lean leader to keep up with. When the leader recognizes that his or her job is to act as the guarantor of the integrity of the processes in his or her area, it becomes clear that this is the new work for a Lean leader. There is no delegating the task of personally establishing the importance of process focus in a newly established Lean operation. Over a period of years this kind of follow-up can become less frequent, but it never goes away; it defines a "new normal."

Attribute 8: Effective Relations with Support Groups

A Lean operation is a finely balanced system. All of its component parts must operate effectively for the whole to succeed. Manufacturing is an important part in the equation, but not the only part. For example, here are some other considerations:

- What does it matter if manufacturing has reduced the time it takes to change over from making one model to the next if the equipment involved breaks down on a random, frequent basis?
- What benefit is there from a well-balanced, smoothly flowing assembly operation if the purchased parts are constantly out of stock?
- How significant will be the impact on morale from a well-run suggestion system if the safety of those in the area is not given the highest priority?
- What will be the benefit of a system for frequent rotation among workstations if HR policies have not been changed to reflect this requirement?
- What is the point in establishing visual pull systems if accounting still insists on recording transactions every time inventory is moved?

Clearly, support groups—production control, engineering, maintenance, human resources, quality, safety, accounting—need to be part of the picture in a well-functioning Lean operation. The Lean leader needs to think of them as resources, not as convenient repositories of blame for the system's shortcomings. They need to be incorporated into the daily life of the area, involved in problem solving and improvement plans and activities. A diversity of perspectives and resources is needed for a finely tuned Lean system to run smoothly and recover from interruptions. Support groups can bring much of both to a Lean operation.

Not only should these contributions by support groups be welcomed by the Lean manufacturing leader, but he or she should also be able to expect them. After all, it is a fair question to ask: What better way for support groups to deploy their resources than in support of the Lean production process? For this reason, most support groups are expected at the daily value stream meetings, and expected to carry out tasks assigned to them by the Lean leader in response to problems reflected in the performance tracking charts or opportunities for improvement.

Manufacturing and the support groups in a Lean operation need each other too much and too intensely to be able to afford the finger pointing and isolationism that were possible in a batch-and-queue world. Typically, it behooves manufacturing to take the first steps, inviting the support groups into the Lean circle. It is not unusual for manufacturing to be asked to follow up with training in Lean tools and techniques for the support groups, since it is also not unusual for manufacturing to have started off on the Lean journey alone.

Attribute 9: Don't Confuse Measures of Process with Measures of Results

Leading a Lean conversion project is in many ways an exercise in holding team members accountable for producing the results as documented and anticipated in the project plan. Good project management practices include frequently verifying that intermediate step-by-step tasks have been completed. When small steps run behind schedule, the project manager has the information to intervene in the early stages of a developing problem. The goal is to head off problems when they are still small, preventing them from growing large enough to threaten overall project schedule and budget performance.

In the world of project management, close monitoring of day-to-day, step-by-step progress is the equivalent of Lean's emphasis on frequent focus on process. Those new to Lean are often shocked at the emphasis on bringing problems to light, compared to the common conventional practice of covering problems and burying the explanation. Lean thinkers, as Shingo observed, regard interruptions and breakdowns as nuggets to be mined for clues to the best next places to focus on improvement. Put another way: take care of your process, and your process will take care of you and produce the results you expect.

In Case Study 9.3, it took me a few days to understand fully what the engineer and I had seen and heard. I had learned the manufacturing VP who recently became responsible for this business unit was a classic results-oriented manager. His leadership style was described as direct, blunt, and demanding. He set the goals. The plants submitted their plans. He approved them, but only after reducing proposed budgets and upping projected results.

CASE STUDY 9.3: RESULTS-ONLY FOCUS CORRUPTS PROCESS MEASURES

In a visit to a plant said to be far along the Lean and Lean management journey, I was surprised to see "pitch attainment" on the agenda for its twice-daily production meetings. I was more surprised to learn the goal for this measure was 95 percent for each of the plant's four value streams. Pitch for each value stream was set at 30 minutes; that is, the actual versus expected number of units produced was measured every 30 minutes. For an eight-hour shift, 95 percent meant 15 of 16 pitches meeting expected production.

Going into this plant, my understanding of pitch measurement was straightforward. Start with a production tracking chart based on takt time or another measure for expected pace of production. Using the tracking chart, monitor production and record the health of the process by documenting each instance of problem, such as flow interruptions, process misses, and system breakdowns. Then, through a standard accountability process, assign tasks first to understand the root cause of the problems, and then to eliminate them. (In this plant, the only production visual controls were in assembly areas, themselves a puzzlingly incomplete implementation of visuals.)

So, the expected target of 95 percent pitch, and the label "pitch attainment" seemed odd to me. I gemba walked later that day with an engineer from the plant Lean team. I was interested in the visual controls the plant was using, so we looked closely at value stream information boards, then moved on to the production tracking chart in the assembly area of one of the value streams.

This was a mixed-model value stream producing two kinds of units, freestanding and wall mounted. Takt time and pitch were the same for each type of unit: 60-second takt and 30-minute pitch, for 30 units every 30 minutes. Assembly alternated every 30 minutes between freestanding and wall-mounted units. The production tracking chart had a row for each pitch, for example, from 7:00 to 7:30 a.m., and columns for expected number of units, actual number produced, variation from expected, cumulative production, cumulative variation, and reasons for misses. If actual number of units fell short of expected, the actual number produced was highlighted in red. When actual met expected, the number was highlighted in green.

The tracking chart was pretty standard, but what was recorded on it stopped the engineer and me in our tracks. After actual production of 30 units in each pitch for the first four pitches of the day, the rest of the pitches alternated between 30 and 28 units completed. Yet, the actual of 28, short of expected production by 2 units, was highlighted in green with nothing written in the "reasons for miss" column on the chart. As the engineer and I stood in front of the chart scratching our heads, the assembly team leader came up, introduced herself, and politely asked if we had any questions.

We told her we could not understand why pitches that were short two units still showed green with no recorded reasons for misses.

Here is what she told us: "We've been short parts for wall-mounted units, but the whole team's here today and the line's been running good, right at takt. If we'd of had the parts, we would've run 'em at takt, so I colored those pitches green."

Then it was up to the plants to perform. At first, when walking through the plants occasionally, this executive saw red entries on production tracking charts. This plant had been using the same kind of tracking for several years. Red meant missed pitches in assembly or, in upstream areas, problems with equipment, material, or yield. Red meant problems. Problems threatened results. Problems, therefore, were not good. Problems attracted attention—and heat—from the VP. When all measures were green, he shifted his attention to someone else's plant.

I knew the plant manager and his staff well enough to trust that nobody had been told to cheat on their performance measures. Yet, I also knew in the chain-of-command organizations in which most of us work, it is important to deliver what the boss asks. In this case, the boss was asking for green pitches 95 percent of the time. Altogether, the message was clear enough down to the level of team leader. Fudge, tweak, or put your own interpretation on the measures if you must, but make them hit the target. This turns out to be a common experience, one you may have heard about or seen.

Lean designs are finely tuned and delicately balanced, in part to show quickly where abnormalities interrupt production or cause defects. Focusing on process, a principal outcome of Lean management, can easily be subverted by the "more is better" thinking characteristic of a conventional results orientation, as in: if some green pitches are good, more green is better, and all green is perfection. When a conventional thinking leader sees red on a tracking chart, he or she may figuratively be "seeing red." That is, "If our Lean conversion is a success, why am I seeing red pitches? Green is our goal, so green is what I want to see!" You can practically hear the words, right?

Measuring percent pitch attainment, the result of this thinking, corrupts pitch-by-pitch performance as a measure of process health. Calling it pitch attainment causes this classic process measure to morph into yet another end-of-day tally of results. Habits die hard; left unchecked, the results habit drags process measures into yet another backward-looking measure. Your process has almost certainly not attained perfection, but compromised measurement suggests it has. With that, conventional results-oriented attention shifts to the next crisis. In doing so, the conventional leader turns away from the search for improvement, away from what is truly at the heart of Lean.

A Measure of Process or Result?

Many subtle distinctions separate Lean thinking from a conventional batch-and-queue approach. One of them involves measuring process versus measuring results. It can be tricky. How can you tell when a process measure is no longer being used to reflect the health of the process? What marks a process measure that has been corrupted by the habits of results thinking—even if inadvertently?

One clue is batch size. Here, batch refers to the length of time reflected by the measure, for example, an hour, a day, or a week. The Lean adage "Smaller quantities more frequently!" reminds us to be wary of large batches and applies to measures just as surely as to inventory. Effective process measurements occur relatively frequently, with relatively brief intervals between observations. The decision you make about how often to take measurements answers the question: "How long do I want to wait to know whether my process is operating normally?" The Lean thinker's answer is: "Not very long!" (though see "Frequency of Observation," below).

Measurement that frequently records performance is designed to catch process misses, interruptions, abnormalities, or more generally, to discover problems quickly and close to their source. Frequently checking how a production process is operating as it cycles exposes problems and allows you to contain them. Frequent measurement also keeps you close in time to the appearance of problems, which puts you in a better position for problem solving when you return to the issue.

By comparison, when considering daily pitch attainment percentage, the time period in question is usually a shift, which batches the pitch-by-pitch results. This is the same effect as reporting an average for the shift rather than showing an hour-by-hour or pitch-by-pitch performance profile. In other words, this kind of reporting, even though *based on* frequent measurement, has the effect of lumping the data points together, smoothing over problems and making it more difficult to see them clearly. As with any other example of batching, process problems are hidden in the batch. The problems could be bad parts, late inpatient discharges, inaccurate specifications, delayed procedures, incomplete information, incorrectly processed orders, late drawings, or pitch cycles with process misses. The problems are concealed by the batch, rather than highlighted unit by unit, case by case, order by order, patient by patient so problems are made visible.

This kind of averaging or summary percentage reporting measures results. "What was our pitch attainment today?" is equivalent to asking: "What was our productivity today?" The point is not that measuring results is bad—not at all. Results are important, but I am sure your organization already has enough measures of results. By aggregating process data across periods, you lose the benefit of process measurement, which is its sensitivity to identifying process problems.

So, if it looks like a process measure but you are not using it to highlight problems, you are missing the point. You already have summary measures of safety, quality, speed, and cost to measure the results of your process at the end of the day. Use them for that purpose. Meanwhile, preserve the focus on specific pitch-by-pitch process misses. Treat these problems as opportunities to improve and make a point of acting on them. Keep a list, a rogues gallery, of the problems discovered and tally the problems permanently eliminated. This kind of scoreboard reflects an emphasis on process, a focus on improvement, and using process measures appropriately, to drive improvement.

Frequency of Observation

Batch size, expressed as the time interval between taking process measurements, is one way to assess whether your measures are process or results focused. Use your judgment when setting the interval for measurement frequency. In a mature Lean high-volume, repetitive manufacturing operation, pitch intervals might be as brief as five or ten minutes, usually determined by pack-out container quantities.

Circumstances are different with lower-volume, high-cycle-time production. It might take several hours, a day, or more to produce a unit when the work is production of technical design specifications or drawings, for example. In a case like this, you might be well served by a measurement pitch of once or twice a day. In production with very long cycle times, for example, refurbishing and upgrading complex electronic gear, the complete job might be measured in weeks. In cases like these, do you want to wait days, or even weeks, to know work is progressing normally?

Probably not. Instead, consider breaking these long jobs into shorter intervals—sets of tasks of a few hours duration. By pacing work this way, you will be much closer to its actual progression and more likely to spot and preserve information about problems, all the better to eliminate them and improve your process. Workers might protest being micromanaged, but as with Case Study 4.4, the point is to focus on the process, not the worker, to expose and eliminate the kinds of problems that interrupt and frustrate those doing the work.

Lean Leaders Recognize Imperfection

Leaders who have learned to be Lean thinkers ask for results, certainly. Results matter, but not above all. Lean thinking leaders also ask: "How many problems have we found? What have we learned about their causes? Have we eliminated the causes? How can we be sure they have not returned?"

Lean thinking leaders know their processes are not perfect, and so look for ways to stress a process that is green *too much* of the time. One rule of thumb is: when measures are green more than 75 percent of the time, lower the water. Shortening lead times, linking processes, cross-training people, decreasing inventory, increasing changeovers—all these stress the system. These are examples of the Lean adage to "lower the water to expose the rocks." Then, go after the rocks, those causes of red measures that are exposed by the stress. Lean thinking leaders know how to mine the rocks for the nuggets they contain, and to convert them to improvement. Eliminate the rocks and the system stabilizes at a new, higher level of performance. That is one route to "reaching the next level."

So, respect the integrity and intent of your Lean measures as indicators of the health of your process. You already have plenty of ways to measure results. Do not compromise your measures of process; pay attention to them and they will help you drive improvement.

Summary: Consistent Leadership Is the Crucial Ingredient in Lean Operations

Changing from batch production to Lean is difficult because the two are so different. Lean management calls for responses that are opposite those that have been learned and reinforced over a career in batch-and-queue operations. No change is easy, but swapping one habit for its contradictory opposite is particularly challenging. Because of that, leadership and leaders' persistence and consistency over time is essential for successfully transforming from batch to Lean.

George Koenigsaecker, one of the earliest postwar Lean practitioners in the U.S, has noted that leadership is always in short supply, which may account for the disappointing track record of attempted Lean conversions. A slightly different slant on Koenigsaecker's idea is from Rensis Likert, a notable management theorist and researcher who noted, "Nothing changes until leader behavior changes." That is the point of this chapter; what leaders learn to do is the crucial ingredient in a successful Lean conversion.

Specifically, a leader's behavior is more important than his or her personality—being inspiring, charismatic, or emotionally impactful. I have seen forceful, charismatic, inspiring leaders utterly fail to understand what was needed to sustain Lean conversion efforts. I have seen leaders with a good conceptual understanding of the need for Lean management utterly fail to execute the basics, such as teaching subordinates and holding them accountable for applying what they had learned. The leaders of successful Lean transformations, in my experience, were relentless in teaching and expecting to see live, on-the-floor examples of Lean management in large and small examples everywhere. You would call them dogged in their persistence; you would call few of them charismatic. Significantly, none of them knew anything about Lean when they started; first and foremost, they were both teachers and students, and have remained so.

Success in a Lean conversion ends up depending on those who lead the organization. They must teach, inspect for, reinforce, and hold all accountable for management practices consistent with the principles of Lean. As leaders improve their Lean management skills, their newly Lean operations stabilize and begin paying off in better and better performance.

Nobody is born knowing these principles and how to implement them. Everyone has to learn them through practice, trial and error, and coaching. You can learn them, too. That is one of the main points of this book. With exposure to a few straightforward principles and some examples of how they can be applied, you can learn to lead and implement Lean management and, by doing so, see your Lean conversion live up to its potential.

Study Questions

1. What is the case for change for Lean here? Do you and others find it inspiring, or credible, or motivating? What might improve it?
2. How would you describe follow-up here? Disciplined, haphazard, sometimes inconsistent, variable from leader to leader, situation to situation? What is the impact on process improvement?
3. How are problems treated here? As things to be handled for the moment, issues to be covered so we can move on, or opportunities (however momentarily unwelcome) to make improvement?
4. Does new process design here take account of differences that will be called for in leaders' behavior in support of the new process? What effect has this had on the success of changes in processes?
5. Is there some balance between the focus on process and the focus on results? How does this show in what things are measured and what stimulates action in response?

Chapter 10

Solving Problems and Improving Processes—Rapidly

The Lean management system consists of more than its four principal elements; a root cause orientation to problem solving, a rapid response system, and a progressive approach to process improvement are among the elements that round out the system. This chapter covers each of these topics. (Chapter 11 covers people-related elements.)

A Root Cause Orientation to Problem Solving

Any production process, whether manufacturing, healthcare, or administrative, will experience problems that interrupt or threaten production. That is as true of Lean systems as of any other. A key distinction between Lean and batch-and-queue systems lies in how the inevitable problems are dealt with by frontline people and leaders.

The typical response in batch and queue is to work around the problem, often in creative, sometimes unconventional ways. The focus, of course, is on doing what it takes to meet the schedule. In a Lean environment, the typical response is to ask why the problem occurred and what caused it.

This reveals one of the least obvious aspects of the Lean production and management systems: Lean is fundamentally an improvement system. Implementing a Lean production system does not solve problems. Instead, Lean exposes problems, so you can see them, define them, and eliminate their causes, and improve.

You have probably heard the metaphor about implementing Lean: you are "lowering the water to expose the rocks." As has been repeated here, as you implement Lean production, you reduce the extra resources—both formal and informal—that were in place to protect the batch-and-queue system from long-unaddressed problems. These "just in case" resources are the water that Lean implementations lower. The idea in Lean is to systematically expose the problems that have always been present, but covered over, hidden by the extra resources, like the water behind a dam covers what used to be dry land. Once exposed, the problems can be analyzed for cause, and then eliminated.

As you lower the water, the problems show themselves. Some are listed below:

■ Setups take hours and hours. Batch solution: Keep extra inventory on hand to cover the changeover time.

■ Planned patient discharge times are rarely met, tying up beds. Patients are admitted through the ER when there is no inpatient bed available. Batch solution: Admit patients from the ER regardless, park them (literally) in ER corridors, and designate a "hallway nurse" to look after them.

■ A work center produces defects seemingly at random a quarter of the time. Batch solution: Keep extra inventory on hand to draw on in case we hit a run of bad parts.

■ A piece of equipment goes down like a yo-yo, but does not come back nearly that fast. Batch solution: Keep other equipment on hand that in a pinch can be set up to produce the needed parts, even if the process is not quite the same.

■ A rash of special orders comes in that only Lai Yue knows how to process, but it overloads her. Batch solution: Call the customers and push out lead times. Tell Lai Yue she has to work Saturday and Sunday.

■ A supplier just cannot seem to deliver the full order, at least not on time. Batch solution: Again, cover this unreliability with extra inventory to draw on, just in case.

■ An employee chronically underperforms. Batch solution: "Well, that's just George. He is just like that some days." Work some OT or add a body to the line to take up the slack.

Lean exposes these problems, some of which you knew you had, but had given up ever resolving. As a perverse kind of bonus, Lean also exposes problems you had no idea were lurking, waiting for the chance to bite you. Yes, that really is the sound of opportunity knocking; it just sounds like you gritting your teeth!

Workarounds Are Anti-Improvement

Once problems present themselves in the form of flow interrupters in a variety of disguises (process quality, equipment reliability, supplier performance, support group responsiveness, operator skill deficits, leadership commitment, overly long lead times, late shipments to customers, insufficient capacity, leased space to hold safety stock inventory, crowded waiting rooms, and so on), you have two choices.

The first is tried and true: go back to the stashed extras you were supposed to have gotten rid of with the Lean implementation but had not quite gotten around to. Move them into the process. Do a workaround and meet the schedule. The schedule is the object, right? Tomorrow maybe things will work out better.

The second alternative is to do what you must to patch up the process. After all, you have customer commitments to meet. But as you do, take careful note of what seems to have gone wrong. Document it in the management system tools you are beginning to use. Come back to the problem as soon as you can, and begin asking simple questions such as: Why? What was the cause of this interruption? Drill down to the root cause. Identify the top three interrupters for the week or month. Begin to work on either preventing recurrences or setting alarms to alert you that an episode of a known problem is about to happen. Keep working progressively down the list of known interrupters as you discover and eliminate root causes. Pick the easier problems first as you practice and learn problem solving.

A New Way of Thinking

This sounds simple, but it takes a new way of thinking from the perspective of most who grew up in a batch environment. Consider this contrast: in batch-and-queue systems, leaders are expected to work around problems quickly to meet the schedule. When the workaround is successfully in place, the problem is considered solved; tomorrow is another day. By comparison, Lean leaders are expected to perform cause analysis and put a root cause solution into effect (Case Study 10.1).

CASE STUDY 10.1: NEW LEADERS
UNCOVER THE CAUSE OF OLD PROBLEMS

A case illustrating this difference in thinking comes from an area building a hot new product. The product included upholstered components that were subassembled in a process for sewing and upholstery linked to the final assembly line and supplying it in sequence one unit at a time. The assembly line team leader was scrupulous about maintaining documentation of flow interrupters that caused his takt-paced line to miss its goals for hourly performance, even though his records hardly ever sparked action.

A common flow interruption at final assembly was wrinkled upholstery. When that occurred, the unit had to be shunted to the repair area, disassembled, the upholstered subassembly returned to the feeder process, taken apart, and reworked, and the unit reassembled. This process caused considerable delay; the result was missed output goals at final assembly. Initially when these problems occurred, they were simply reworked, often by the team leader in either final assembly or the upholstery area. The workarounds caused some overtime, but because it was not tracked separately from the regularly scheduled overtime used to meet strong customer demand, it was not examined.

A retirement and two reassignments brought new leaders to the area. They paid more attention to their resident Lean sensei than had their predecessors. The previous leaders had seemed more interested in maintaining their numbers for cost and schedule completion than in bringing the area to a new level of Lean operation.

As the new leaders in the area progressed in their Lean thinking, they started to analyze the hourly tracking charts from final assembly and the log sheets from the final assembly "hospital," or repair area. The new value stream manager and her two supervisors found upholstery to be the single leading cause, both for units going into the repair area and for the final assembly line missing its hourly goals. Further investigation (including a more specific definition of what was meant by "bad upholstery") identified wrinkles and puckers as the most frequent causes for rejects.

The upholstery supervisor followed up. He found standardized work in the upholstery area called out the correct process, but without much in the way of specific detail. He also found that the upholstery process was rarely monitored for anything other than volume of output.

Further investigation showed that the main causes of rejected, wrinkled upholstery were either not enough fasteners used to attach the fabric to the substrate, uneven spacing of the fasteners, incorrect sequence of fastener application, or all three. In addition, wrinkling varied between two types of upholstery fabric, requiring a different number and spacing of fasteners for the two different classes of material.

The supervisor and team leader in the upholstery area, having reached an understanding of the leading causes of rejected subassemblies, built a life-sized display of the correct number and spacing of fasteners for the two different materials, retrained the operators, and monitored performance against the newly clarified standard. The wrinkled upholstery problems virtually disappeared from final assembly.

In the meantime, a downturn in orders had occurred so that overtime was no longer routine. Now, extra hours were directly traceable to production problems. With the change in the upholstery process, a decline in unplanned overtime could be observed as a result of this improvement.

Should Perfection Be a Goal?

This example is not a dramatic one, but it illustrates an important point noted in Chapter 9: Lean thinking involves being serious about pursuing improvement with the goal of eventually reaching perfection in process and execution. Before the change in leadership in Case Study 10.1, one of the team leaders was usually available to come off his or her job to perform the rework, even though that meant skipping other duties (that potentially could have included monitoring standardized work in the upholstery process). The problems were virtually never severe enough to cause a miss in daily schedule completion. In the previous batch-and-queue environment, one that still existed in other parts of the plant, working some overtime to achieve daily schedule completion was routine and perfectly acceptable.

Several things made the difference in turning the Lean corner in the upholstery operation, which by the way, went on to many more incremental improvements that, taken together, have transformed the appearance, performance, and productivity of the area. One factor was faithfully maintained documentation at final assembly. When attention shifted from schedule completion to process quality and yield, the initial data were available right away for Pareto analysis to reveal upholstery as the leading source of interruption.

A second factor, and the most important, was a gradual change in the mindset of leadership in the area, prompted by ongoing coaching by the plant's resident Lean sensei. As thinking shifted from schedule completion to process quality and yield, the leaders began to ask why they had continuing upholstery rejects. They took a systematic approach to discovering the causes of the problems, and to eliminating them so the problems did not recur. Once they focused on improvement, rather than living with a long-standing but manageable problem, they were able to make a permanent fix, a permanent process improvement. The supervisor in this area has come to see perfection as a goal, though he does not talk about it in these terms. Rather, he talks about the need to make further improvement. A clue to his new, larger goal of perfection, however distant and even indefinable, is that he sees no end to possible improvements in his area.

Structured Problem-Solving Process

It is one thing to set perfection as a goal; it is another thing to put in place the tools that allow you to make progress toward the goal. A structured problem-solving process is an important step on this journey. At first, this might be no more demanding than asking people to apply the five whys when problems arise. Training and coaching frontline leaders to teach the five whys problem-solving method ensures consistency between what people get trained to do and what their leaders support and reinforce them for doing. More sophisticated approaches can come later. Taking people through the steps, one at a time, with each step closely followed by application, is often the most effective way to introduce these ideas.

Later, or initially, if appropriate to the readiness of the leaders involved, if you are not familiar with one, find a six- (or seven- or eight-) step problem-solving process and learn how to use it. Conventional training closely linked to application can work. Or, you may simply want to start asking for the steps one at a time as circumstances dictate, coaching your leaders on how to carry out each step. Depending on the number of people you need to reach, it might be enough to deliver the training on a just-in-time basis as the need for the next step comes up and a teachable moment presents itself with your leaders and team members, too. The basic steps are the same regardless of the brand of process you use and how you train people in it:

Step 1: Identify and define the problem.
Step 2: Quarantine the problem and take other immediate remedial actions.
Step 3: Involve the appropriate, knowledgeable people.
Step 4: Conduct root cause analysis.
Step 5: Identify root cause solutions, assess them, and test the preferred alternative.
Step 6: Implement the root cause solution.
Step 7: Monitor and revise the solution as indicated by performance data.

Who Makes Improvements?

The orientation to cause analysis and pressing for perfection one step at a time will quickly produce opportunities for improvement that outstrip the capacity of technical-professional support groups. In a conventional batch environment, this would be just another instance of the typical bottleneck in getting engineering, or quality, or production control, or whomever, to work on a request for a process improvement made by a production floor leader. That is because in a conventional organization, improvements are "left to the experts."

In this conventional scenario, a technical project team moves into and then out of a production area, leaving in its wake what are intended to be permanent fixes. Sometimes these fixes only work on paper, in which case they are quickly undone. However, sometimes the project team's fixes work quite well. But, because it works, it gets no further attention. The process remains static, often for many years, without further improvement. No further changes are made. That may be because the leader's mindset ("It ain't broke!") blinds him or her to further opportunities. Or, leaders may not know how to lead improvement efforts themselves with mostly their own resources. The likely waiting time for a response by a technical project team is often so long that it is not worth initiating the request.

Short-, Medium-, and Longer-Term Improvements

In a Lean environment, the expectation is that everyone has two responsibilities. The first is to run your part of the business—even if that is only a single workstation—on a day-to-day basis. The second is to improve the business, or contribute to improving it, *continuously*. We have already seen one example of this dual responsibility in the daily accountability process and its task assignment board. Improvement efforts in Lean management usually fall into one of three categories (with movement between categories as experience dictates) determined by scope, scale, and duration of the improvement task. Table 10.1 lists these categories.

Simple problems or straightforward improvements can be assigned and managed in the daily accountability process three-tier meetings. For example, a team leader might be assigned to make a shadow board for tools currently kept in drawers and often missing when needed. Or, he or she might be assigned to create labels for cabinet doors and shelves identifying the materials and supplies stored there. With the shadow board or labels in place, the team leader might be assigned to create a simple daily audit system to sustain adherence to "a place for everything and everything in its place." The next assignment might be to update team leader standard work with items like these:

■ At end of shift, verify tools are back in indicated places.
■ Verify materials and supplies are where indicated and replenished as needed.

Table 10.1 Options for Improvement Activities

Duration of Tasks	Typical Focus	How Managed
1–5 days	Fix an immediate problem, simple cause analysis, implement a simple improvement	Daily task assignment board; follow up at three-tier meetings
6–30 days	Problem-solving process for more complex cause analysis, solution, or recommendation	Via one-page (A3) visual project plan reviewed at weekly project review session
30–90 days	Longer-term or more complex problems or opportunities	Via one-page (A3) visual project plan reviewed at weekly project review session

Note: While these categories do not explicitly include acting on employee suggestions for improvement, any such suggestion can be easily fitted into one of these three. Chapter 11 covers employee improvement suggestion systems in detail.

Recommending Future Improvements

Another outcome of an initial investigation from a daily task assignment is to recommend a project for the list of future kaizen improvement events in an area. For example, investigation of the cause of an intermittent backup on an assembly line identified interruptions in a packing operation caused by poorly organized parts and materials that often resulted in the operator having to leave the workstation to search out the proper materials. In this situation, two kaizens in succession have attacked organization, material resupply, and parts presentation in the pack area. The kaizens were one week in duration, held about six weeks apart. The second event refined the work of the first based on the intervening experience. Again, it is frequently the case that solving one problem reveals others, and more opportunity to improve!

Managing Improvement Activities

You may, at first, find it confusing to determine the relationship among short-, medium-, and longer-term improvements (Case Study 10.2). The best advice is to start, be systematic, and sort things out as you go. Experience says you will figure out a method that makes sense for you and those who are involved in it. Do not hesitate to consider these arrangements as experiments on the way to establishing a stable process. Here are a few guidelines to use in sorting out your process:

- Make the accountability process the responsibility of a specific individual or position.
- Document the accountability process with an agenda for consistency from person to person or day to day.
- Assign each improvement task to a single person.
- Manage improvements visually in only one location.
- Manage each improvement with only one schedule.
- Hold review sessions at least once a week.
- Avoid projects that extend more than 90 days; create a series of 60- to 90-day subprojects as needed.
- Avoid making lists (black holes) of projects; manage the queue visually.
- Group activities of like duration with like management, e.g., daily board, weekly A3s.

CASE STUDY 10.2: LONG-TERM SYSTEMATIC CAUSE ANALYSIS OF AN ASSEMBLY LINE PROBLEM

Some problems are more complicated; for example, if laminate tops are being scrapped because of units moving intermittently on the bed of a shaping machine, is it the machine, the shaping tools, the material, the method, the operator, or Mother Nature? In this case, it was not at all clear; each potential cause had to be ruled out by a problem-solving team that was formed to investigate after several fixes had been attempted, only to see the problem recur. (It turned out to be Mother Nature in the form of seasonally varying humidity, to be brought under control by erecting a humidity-controlled holding area for the substrate material.) This kind of project required systematic cause analysis involving several people, including technical specialists, operators, team leaders, and the supervisor. The team reached its conclusion in the 30-day period and made its recommendation.

This is an example of a problem-solving project readily managed on a generic visual project form. Toyota (and now many others) refers to this form as one version of an A3, after the technical name of the tabloid-sized (11 × 17 inch) paper on which it is printed (see Figures 10.1a and 10.1b). The summary-level Gantt chart on the A3 project plan makes it easy to color-code completion of milestones (green in the week due for on time, and red until the week completed for overdue milestones), so it lends itself well to visual management of expected versus actual. Projects of greater scope, duration, or complexity might be managed with more sophisticated project management tools, but a project manager would use these in the background. In such cases, the A3 serves as a summary visual control for planned versus actual accountability for the weekly pace of progress on the plan.

In this example, the recommendation of the problem-solving team turned into a project to design, specify, procure, install, and test the humidity-controlled holding area. This is one example of a project of between 30 and 90 days duration, also readily manageable in a weekly project review in front of a board displaying the A3 project plans and green or red status of that week's milestones. Some of the project tasks might have shown up on a daily task accountability board. An example might be for the supervisor to review a mock-up of the holding area with the out-cycle operators who would be involved in retrieving the loads of substrate.

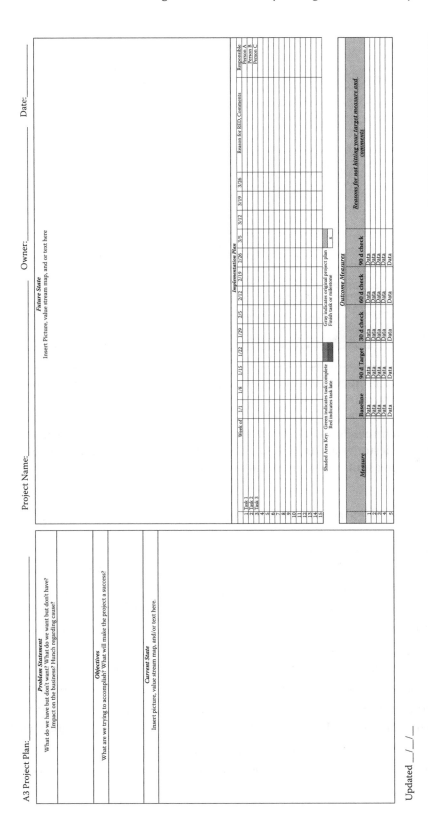

Figure 10.1a   A3 project plan form.

Figure 10.1b   A-3 project plans of a project plan board. Rows one through three on this project board are for each value stream in this plant. Row four is for plant projects. Active projects are in columns one and two, on-deck projects and kaizen plans in Column three. Plans are reviewed weekly. Weekly milestone due dates are noted in gray. Milestones are coded green when completed, red until completed. A note in the blank spot, upper right, says two on-deck kaizen plans required.

Improvement Resources and Skills

One of the roles of a site Lean leader is to teach floor leaders how to recognize opportunities for improvement, and then how to act on them with their own resources. A frequent approach is for the site Lean leader to teach and coach supervisors and team leaders to lead kaizen events. These are tightly focused, time-limited (usually no more than a week, often less), small-scale intensive improvement projects.

Kaizens are specifically designed to make incremental changes, and there may be a series of kaizen improvement events that focus repeatedly on a single small area. The results are cumulative, gradually making step after step in improving flow of production, easing the physical tasks of the operators, reducing setup times and inventory, improving quality, making equipment more reliable or available, and so on. An additional important feature of kaizen events is their makeup. The composition of the kaizen team, usually seven or fewer participants plus a leader, is mostly frontline people from various areas in the facility. Participating on a kaizen team is a very powerful method for members of the frontline workforce to learn the principles of Lean.

Technical professionals will often provide consultation and support to kaizen teams along with the site Lean leader, but most of the work is done in the teams themselves, which is one of the reasons kaizens are such powerful vehicles for training. Concept meets application in a kaizen team, and the application is direct, hands-on experience.

A Rapid Response System and Implications for Support Groups

You have team leaders on the floor paying close attention to the way the process is operating. The new Lean approach is great when it works, but quickly comes to a halt when things go wrong. Ouch! When that happens, you need a response—fast. There are several ways to summon help quickly. The range includes simple call lists (be sure a phone is close at hand), specially designated pagers, radios, or cell phones (often red in color) carried by designated support group representatives who are "on call," escalation systems based on number of missed cycles, automated escalation systems, and so on.

Do not be lulled by these arrangements; a rapid response system is not about the cool technology you use to summon assistance (Case Study 10.3). The necessity of changing the connection between support groups and production areas is an often overlooked aspect of Lean production. For Lean to be sustained without yielding to the temptation to go back to buffers of time, labor, and material, quick responses are needed when interruptions arise, as they definitely will. What comes to the surface is simply this: the most important activity in a production facility is *production*. The tasks and projects that engineering, programming, skilled trades, quality, supply chain, and scheduling work on are not unimportant. They are just not *as* important as what happens on the production or service delivery floor when one of these specialists is needed to, as the name indicates, support production.

Support Group Response Time

This is often a new idea. In the previous batch-and-queue operating environment, there were buffers of hours or even days between operations. When one element of the process experienced problems, it was naturally isolated from the others by the inventory waiting between each process—the "queue" in batch and queue. In this world, it was important to respond to problems on the floor, but rarely was there a sense of urgency to the response. Support groups had goals and projects that they focused on. The phone calls (or emails) could usually wait, and usually did, but no more, not in a well-disciplined Lean implementation (Case Study 10.3 and Case Study 10.4).

Some might immediately conclude that a different organization structure is required, that is, a value stream organization in which support group members report directly to the value stream manager. Depending on the disposition of higher-level leaders, this can become a political struggle that absorbs time and energy and does not always yield the hoped for outcome. An alternative is to focus on the production process and the accountability process that helps to sustain it. Case Studies 10.3 and 10.4 describe two scenarios where the presence or absence of accountability made all the difference.

It is typically easier to gain commitment for measured, timely response to a call for support than it is to get the organization chart changed. The former is like motherhood and apple pie: Who can argue with agreeing to a rule-governed quick response to a stoppage in production or a customer problem? On the other hand, a proposal to change reporting relationships can easily be perceived, and often rightly so, as an assault on a fellow manager's power base. You have enough things in your newly converted Lean process to worry about already without picking that kind of fight (Case Study 10.5).

Leadership Alignment versus Changing Reporting Relationships

These case studies illustrate the often unanticipated implications for the organization that follow from the newly Lean production area. For Lean management to work effectively, the support groups have to be prepared to respond at a pace determined by the takt pace of production—at a minimum, that means support groups realigning priorities and leadership expecting accountability for the new priorities. Organization structure is

CASE STUDY 10.3: WORKING WITH A RESPONSE SYSTEM (NUMBER 1)

The first case occurred, not coincidentally, in the plant where the supervisor "stored" the production tracking forms on the floor under his desk. The Lean conversion involved switching from a build-complete process at a number of stand-alone two-person workbenches to a single progressive build assembly line. In the bench build process, a number of subassemblies were produced in batches upstream from assembly, set up in kits based on a schedule, and pushed up to the benches. The progressive build line incorporated a number of these subassembly areas in an integrated takt-paced flow production process. Signals went out to subassembly production from the single schedule point in the flow process.

It was extremely rare for production to stop in the bench build scenario. There was always the option of working ahead based on the schedule. And, if one bench experienced problems, it was likely the others could keep on producing while the problem load was pushed aside and the next one brought in. In the flow process, when there was an interruption, 13 people stood around waiting for the problem to be identified and fixed. Two temperamental pieces of automated equipment were built into the flow process. They failed repeatedly. Worse, the failures were intermittent.

To the credit of the project team designing the new flow process, they recognized the need to establish a quick response process and told the support groups they would need their support in a different and more timely way than in the past. Maintenance, engineering, and materials management agreed, and for the first several weeks when the "911" calls went out, the response was timely. A host of flow interruptions quickly emerged as the new process began operation. There were frequent material shortages outright, and material supply errors resulting in mix-ups in the sequence of parts to the various stations on the progressive assembly line. There were problems with alignment of parts in assembly that used to be muscled through at the benches. Now, these problems caused repeated disruptions in flow at the 36-second takt time. The automated equipment also failed intermittently, causing further interruptions.

It was a pretty typical start-up for a Lean conversion. The system was doing what it was supposed to do, exposing flow interruptions. Unfortunately, there was no accountability process in place.

The increasingly frustrated value stream manager reported the maintenance, fit-up, and equipment problems in daily plant staff meetings, the way it had always been done. The response was to put the problems on a list of functional projects, also the way it had always been done. The response system also quickly regressed to operating the way it always had, slow and uneven.

The plant manager at the time had left the Lean conversion to his project team. The new layout looked good from the aisle, and when everything was right, it ran beautifully. But as soon as the project team, with its support group specialists, disbanded a few weeks after the start of production, there were no extra resources on hand to resolve problems that repeatedly surfaced. The plant manager expected nothing different in the way his support groups responded. And he had no inclination to change the reporting relationships of the support groups. The management system had not changed in the rest of the plant, and the converted area continued to struggle until the product line was moved out of the plant.

CASE STUDY 10.4: WORKING WITH A RESPONSE SYSTEM (NUMBER 2)

The second case involved the same plant as the first, and two of the people who had served on the project team for the first project. One of them had been the co-leader responsible for the Lean technical design alongside the person who had become the value stream manager. In the new project, this former co-leader wore both hats, project team leader and value stream manager. Two other factors came into play. Instead of converting an existing batch-built process to a Lean one, this project involved a high-profile new product. And there was a different plant manager. The organization structure and reporting relationships of the support groups remained exactly the same as before.

This time, the value stream manager implemented a complete Lean management system along with the physical implementation. Because it was a new product, the plant manager's boss was more engaged in the project, and so was the plant manager. The company had had several years of experience in Lean conversions since the first case. More people had been exposed to the tight interdependency within a Lean production process, as well as between the production area and its support groups. The plant manager agreed to be part of the response system, the last one on the call list when production was stopped. According to the response system, he was to be called when production was interrupted for half an hour.

An element in the response system was a log sheet kept right by the line-side telephone from which the 911 calls were placed. The log recorded actual versus expected response times in the escalating network of calls, and reasons for misses when they experienced delayed responses to calls for help. These reasons for misses were reviewed no later than the next day.

There were few misses on the occasions when 911 calls went out. The supervisor virtually always responded within the expected three minutes. The plant manager always responded when his number was called, even if the response had to be by phone. He expected a timely, in-person response from the support group members assigned on a given day to wear the response system red pager.

This time, the response system worked. Leaders were aligned and the management system was in place to record, review, and act when actual did not meet expected throughout the process. No change in organizaion structure was needed, but for the system to work as it did, the many other changes noted above were required.

CASE STUDY 10.5: EXPERIMENT WITH CHANGING PROCESS BEFORE STRUCTURE

A small engineering group was responsible for designing a type of mission-critical, custom-designed hardware for naval warships. The units designed by the group had been installed in many ships over a number of years; the installed base was large. The group's engineers also responded to requests from the field for technical support on the in-service units it had designed. The requests came in via telephone to a support person and were directed to the engineer responsible for the in-service unit with a particular serial number.

Responding to these field support calls was a responsibility in addition to the design engineer work of each engineer. The calls interrupted the engineers' concentration on the detailed design work of these high-precision, engineered-to-order units. With many units in service, the calls had become frequent enough to be a major disruption in the group. The result was an increasingly serious problem in timely completion of work on new units.

The engineering manager did not know what to do about the problem. He had attempted to get approval for adding engineers to the group to increase capacity, and tried to shift the field support duty to other departments, but to no avail. He could not get budget approval for any new positions, and the customer insisted that the design engineering group, who best knew the product, should be the ones to respond to calls for support.

The root cause of the problem seemed to be interruption of the in-cycle work, designing new units, by out-of-cycle work, responding to field support phone calls. The proverbial lightbulb went on for the engineering manager when we discussed this possibility. Further conversation led him to the conclusion that he could isolate the out-of-cycle calls, insulating the in-cycle work from interruption without farming out the field support work to a different group.

He decided to experiment with establishing a phone duty rotation among the engineers. All field support calls were routed to a single engineer during his or her assigned daily hour of phone duty. That engineer would take the calls, handle them where that was appropriate, or set a callback time at the top of the next hour when the customer could expect a reply. The phone duty engineer would email a callback appointment with a brief summary of the issues to the engineer returning the call.

The engineering manager later reported that the experiment had been a success. The timing of callback appointments had needed some adjusting, but as it turned out, most of the calls were not deeply technical, and could be handled by the phone duty engineer. Overall, the engineers' morale improved, the group's productivity and on-time performance improved, the customer remained satisfied, and the manager had no further need to request more staff or hand off the support work to another group. The Lean principle of separating in-cycle from out-of-cycle work made it easy to see and test a process solution that had been hidden by mixing the two kinds of work. There was no need to try to juggle responsibilities and reporting relationships across the organization chart.

not much of an issue when site managers understand the implications of Lean and are willing to hold their staffs accountable for supporting a new, Lean direction.

Organizations need functional groups to maintain ready technical expertise and skills. Nothing in Lean management suggests disbanding the central functional organizations. In the longer run, it does make sense to dedicate support group staffers to Lean value streams on a long-term rotational basis. As the process changes and support relationships in Lean conversions become established, a change to dedicated value stream organizations can become less of an emotional issue than at the outset. On the other hand, as the process changes and support relationships mature, organizational changes to support them may, and often do, turn out to be unnecessary.

Summary: Finding the Root Cause of Problems Is Key

Lean is an improvement system in which several apparently contradictory elements are joined. First, the system is designed to expose problems, occasionally bringing production to a complete halt until the problem is resolved. Second, the orientation to root cause problem solving seems to suggest dramatic, large-scale projects to eliminate problems. Yet a primary method for eliminating causes of problems is to make repeated incremental improvements, clarifying problems and recommending solutions that can often be implemented in a week by a kaizen team. Smaller incremental improvements can be made even quicker by one of a few people in the area working on a less formal "quick win" suggestion in the context of a defined quick win process. One of the keys to effective process improvement is developing a clear understanding of the cause of the problem by using the tools and logic of systematic problem-solving processes scaled to the problem, ranging from five whys to more sophisticated approaches.

Separating improvement activities into short, medium, and longer term is a way to give appropriate emphasis and attention to each kind of improvement.

Not all problems can be resolved right away. Some require emergency or short-term countermeasures that allow production to continue while the cause of the problem is diagnosed. Because Lean production systems are so tightly interdependent, when a problem occurs in production, a quick response is essential. For these reasons, a response system is an important part of any

Lean system. The most challenging aspect of putting such a system in place is realigning the priorities and perhaps the measures of support groups. Otherwise, the response system exists in name and hardware only.

Study Questions

1. Are problems here treated as opportunities to improve, or just one more thing to tolerate and work around?
2. Do solutions to problems here eliminate the causes or put on a patch to protect you from the problem the next time(s) it happens? If patches, what are some examples?
3. How long does it take here to solve problems, or at least make progress on reducing them; do things happen quickly, or does it takes major effort, project teams, and lots of resources before action happens? Examples?
4. Is there an approach to resolving problems that is used regularly? If so, what is it?
5. What kinds of people get involved in problem solving and process improvement here? Enough so things happen quickly, or so few (often only busy technical-professional specialists) that some problems seem never to get addressed?
6. Is it easy to tell who is responsible for taking a given action to solve a problem or implement a solution? Examples?

Chapter 11

People—Predictable Interruption, Source of Ideas

People and their ideas for improvement are close to the heart of Lean production. People can also seem to be close enough to the neck to cause a pain there. Perhaps that is because of Lean leaders' experience when they go to start the production day only to find one or more people have "called in" as unplanned absences. It is a big deal to be missing a person or two when the day's labor plan has been matched to the rhythm of takt time, or the allotted time per workflow, or the expected number of patients or customers. Without just the right number of people, flow does not flow, pull can deteriorate into stock-outs, and the takt beat is uneven and sporadic.

People issues may not seem to lend themselves to the process-focused comparison of actual and expected. In actual application, the Lean management approach works well with matters of attendance, rotation and staffing, performance issues, and employee involvement in process improvement suggestion systems. This chapter shows how. Learning Lean is a hands-on proposition. Effective Lean training calls for something different than the typical classroom approach. The training group in human resources (HR) can help develop coaching-based training if they understand the need. HR policy issues also come into play when talking about people issues in a Lean conversion. Matters such as job grades and classifications, pay systems, start and break times, job rotation, and layoff policies are likely to need attention in support of a Lean environment.

Whom Do I Expect Today? The Attendance Matrix

The first people-related issue is the most frequent one to arise: Who is here for work today? We know people will be absent; we just do not know who and when. Typical arrangements for handling absenteeism include carrying extra people—as many as 8 to 10 percent seasonally—to call on when people unexpectedly call in to say they will not be at work today. So, there are extra people in the building or on the network. Just try finding them when you come up short! Often it is a time-consuming scramble that ends in frustration for all involved. I cannot get the person I need or have been promised; the person I do get did not want to come and is not trained in the work I need to have done. There is a striking lack of process in many places, just like this scenario. The first question a Lean sensei will ask is: What is the process here? In the case of attendance, there is none. Is an absence process possible?

Many attendance tracking processes are limited to the number not expected at work. That is, we use calendars for the coming year to write in those workers we expected to be off on vacation, in large part, so we do not grant too much vacation in any single week to handle seasonally expected demand. So, the vacation log tells me who will *not* be available.

A different approach is to use an attendance matrix (all the people on the team, by every day of the month, a page per month for the year) filled in to tell me those whom I *can* expect to be at work tomorrow. Entries in the matrix identify:

- Those with planned vacation for the day (usually coded yellow for that person's row for the days of planned vacation)
- Those loaned out to another area or otherwise assigned (for example, to a project), and thus unavailable for work for a period of time (coded blue)
- Those on medical or other leave (coded green)

I should be able to count on everybody else showing up, ready to go. When people call in, they get coded red for the day. If they are late, they are coded half red (Figure 11.1).

Do people dislike being coded red? Sure they do! Do we count on everyone we have planned on to show up in order to have a productive day? Sure we do! Should people be accountable for their presence when the team plans on it? You bet! Toyota is said to hold start-up meetings at the beginning of the shift in large part to tell who has reported for work so plans can

Month:________ Attendance Matrix Area:__________

Name	1	2	3	4	5	6	7	8	9	10	11	12	13	14	15	16	17	18	19	20	21	22	23	24	25	26	27	28	29	30	31
Chris																															
George																															
Giacomo																															
Hamid																															
Jolene																															
Juan																															
Kim																															
Kit																															
Marcus																															
Moon																															
Phan																															
Rosa																															
Shaun																															
Willie																															
	1	2	3	4	5	6	7	8	9	10	11	12	13	14	15	16	17	18	19	20	21	22	23	24	25	26	27	28	29	30	31
	Vacation					Call-in				Late				Medical					Project/Loan												

Note: Absence status is shown by color codes. In this black and white example, hatching and shading represent the colors. One color scheme defines yellow for vacation, red for unplanned absence (call-in) or late, green for medical, blue for project or loaned out.

Figure 11.1 Attendance matrix.

be adjusted as needed. At its assembly plant in Kentucky, employees with perfect attendance for the year are eligible to participate in a raffle where new cars are the grand prizes. Showing up is important in a Lean workplace! Think of the savings associated with not having to deal with as much absenteeism as you do today. Of course, unforeseen things will happen to cause even reliable people to have an unplanned absence on occasion. Still, showing up when planned is important in a Lean environment—for everyone.

Who Starts Where Today? The Labor and Rotation Plan

Job rotation through a home rotation pattern is a common feature in the Lean workplace. Rotation helps prevent ergonomic injury from repetitive motion where this is an issue. It results in a cross-trained workforce with the flexibility to move to any of several jobs as needed. And, it means that many people are looking at each job, making a more fertile field for producing suggestions to improve the job for ease, safety, quality, or efficiency. Or, in an office setting, people might be assigned to handle different duties on a rotating basis. This might be to balance different kinds of tasks, or as with the design engineers in Case Study 10.5 in Chapter 10, to isolate interruptions, whether for handling unplanned breakdowns, requests for technical support or customer service, or being on-call for unplanned urgent needs.

Rotation also requires more work for the team leader, who has to establish quickly who starts where at the beginning of work. Relying on memory is one way. But, can the team leader reliably recall who started where yesterday, or where everyone ended? Probably he or she cannot. What about asking people where they started or finished the day, or the same questions, but for people who are off today? That does not seem like a good plan either. An alternative is a simple set of visual controls that go along with the expected attendance matrix and a qualification matrix. Taken together, these form a suite of tools for labor planning.

Completing the Labor Planning Suite

A labor or assignment and rotation plan is a map that identifies the workstations or the range of assignments in a work area—production workstations, team leaders, water spiders, etc., as well as types of service or on-call roles. In most cases, an abstract schematic that identifies only the workstations or the names of other assignments works best. A label (magnetic is helpful here) for each team member's name goes on the map at the location where that person starts or the day's opening assignments. A rotation schedule (clockwise, zigzag, or a matrix of names and workstations or roles) completes the picture. The attendance matrix shows who is expected to be available for the next shift. It only takes a few minutes at the end of the day to set up the next day's labor and assignment plan, moving the name tags from where they were yesterday. This way, people can quickly find their starting assignment at the shift start-up meeting (see Figure 11.2a and b).

Labor and Rotation Plan: Sled Assembly

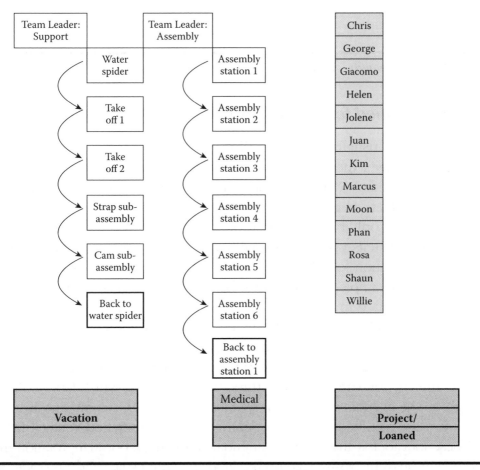

Figure 11.2 **(a) Labor and rotation plan. Column 1 lists support positions. Column 2 shows workstations for (in this case) assembly. Column 3 lists people assigned to this work area or department. The names of people on vacation, medical leave, and Project/Loaned assignments are moved to the applicable list, indicating their unavailability for assembly or support assignment.**

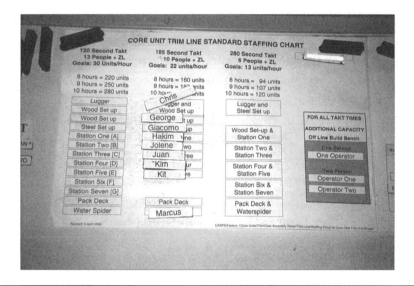

Figure 11.2 (*Continued*)   **(b) Photo of a labor planning board.**

Who Is Qualified for Which Jobs?

Training records usually reside in a file cabinet somewhere, either in a supervisor's drawer (not great) or in the training department or in an HR database file (worse). When someone calls in as an unplanned absence and production is set to begin, you need to know *right now* who you can call on to fill in, even if only briefly to get production going. A qualification matrix (see Figure 11.3) tells you who is qualified at what level for which jobs. It includes information for all the people on your team as well as some from outside it. For example, if others have been interested enough in your area to become qualified in it, or have moved on from your area to another, they would appear on the matrix with the level of qualification they had achieved.

With this information, you are not simply asking for warm bodies to fill in, hoping they can learn the work, keep up, produce good quality, and avoid injury. Instead, you can go to your three-tier meeting, at which labor balance is high on the agenda, and make a specific request for Giacomo and Eva, who you know are qualified to fill the openings you suddenly find yourself facing.

These four tools—the expected attendance matrix, labor plan, rotation map, and qualification matrix—make up the labor planning suite. They provide, at a glance, information about availability, daily starting position, and qualification. Like other visual controls, the labor planning suite raises the level of account-ability, especially the case for attendance with the attendance matrix. The suite also makes patterns visible that may not have been seen as clearly, such as

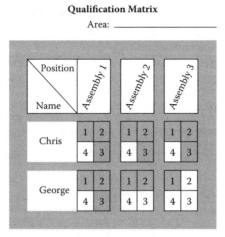

1 = Being trained
2 = Can do the work with assistance
3 = Qualified; can do the work without assistance
4 = Qualified as a trainer

Figure 11.3   Sample skills matrix entries. Gray cells are filled in, representing completion of a given (numbered) level of qualification.

positions where too few people are qualified, the extent of cross-trained people from other departments, or patterns of attendance that had gone unseen.

How Can I Encourage Participation? The Idea System

Setting up conventional employee suggestion programs is quite straight forward. Making them work is another story. Traditional suggestion systems require considerable overhead: engineers costing-out proposed improvements; managers sifting, sorting, and culling; administrative people recording and routing information; and often several months later, the employee being thanked for a suggestion that "we just can't commit resources to at this time." Not very motivating!

When Lean is truly an improvement system, it produces a steady stream of employee-generated suggestions for improvement. The question is how to get the stream started and then, how to keep the ideas flowing. Before an improvement idea system can work, the organization has to want it to work and has to believe employees actually have ideas to contribute and a desire to do so.

And, as I suggested in Chapter 5, the organization must have developed the capability for tapping leaders' latent potential for making bite-sized as well as larger improvements while they also attend to their daily run-the-business tasks. The vacation paradox plays an important role in sustaining process improvement suggestion systems, in the following way.

In a takt-paced Lean production environment or high-volume service operation such as a call center, hospital ER, or urgent care clinic, virtually no time is available in a routine production day for operators or frontline staffers to work on improvement activities outside of structured improvement events, such as kaizens or problem-solving teams. Most of their day is consumed by their standardized work; break time is about all that is left. So, operators are going to have no time to work on implementing improvements they have suggested. And, the benefit from suggestion systems does not come from the suggestions; it comes from implemented improvements. The question becomes: Where do the resources come from to work on implementing employees' suggestions for improvement as an improvement suggestion system is getting underway, and once it is established?

Who Will Work on Suggested Improvements?

The answer is found in the long-term effect of the vacation paradox. Supervisors and team leaders learn through experience with daily task assignments that they really *do* have time most days to work on improvement, especially in a stabilizing Lean environment. This previously unavailable capacity becomes part of the new "way we do things around here." As it does, it becomes possible for team leaders, supervisors, and support group representatives to allocate the time for working on improvement ideas, including those that come from operators through the improvement suggestion system. That is the key that unlocks the gate to sustainable participation in the suggestion process. Consider Case Study 11.1.

A Visual Improvement Suggestion Process

As with much else in a Lean operation, there is power in making the improvement idea process visual. The usual reasons apply: when actual versus expected is visible and followed up, accountability for commitments and performance increases. Posting suggestions for all to see can encourage more suggestions as well as stimulate ideas that build on each other. A visually controlled suggestion process can convert the concept of listening to something you can see.

CASE STUDY 11.1: WHAT HAPPENS
WHEN IDEAS ARE NEGLECTED

This case is a composite portrait that is typical of well-intended Lean implementation projects. When suggestion programs are introduced, especially in the course of a Lean transformation process, many frontline people will submit ideas. Partly, this is a function of the attention the area is getting from the project team working on the Lean implementation. The team will often solicit frontline workers' input on design and feedback on its initial operation. Suddenly, lots of ideas are flowing, because ideas are being listened to and acted on—by the extra resources in the area from the project team.

Then the team begins to pull out and eventually disbands and moves on. They indeed have been able to act on many of the ideas from the area's people who, as a result, typically continue to submit them. The ideas not directly related to the project are often left on a to-do list. And, new ideas continue to come in as people gain experience with the new process. The poor supervisor is left with a pile of suggestions to go along with an entirely new production system to debug and learn how to run. Figuring out the newly redesigned area is where the supervisor puts his or her attention, generally leaving the pile of ideas untouched. The pace of work on ideas slows dramatically and typically stops altogether. At the same time, the stream of ideas is drying up and stopping.

This is usually a frustrating mystery to the leaders in the area, who often genuinely want the help and support of frontline people to make the area successful. The leaders have seen the quality of the suggestions and the lift people experience from seeing them implemented. And now, nothing! But much else is pressing, and soon the leaders' attention has understandably shifted to things about which they know how to do something.

APPLYING THE VACATION PARADOX
TO IMPLEMENTING SUGGESTIONS

The second illustration involves a case of waiting for the vacation paradox to take hold, and then applying it to an improvement suggestion process. This is an example of dramatic change and improvement from the first blush of a Lean transformation in an assembly area. Management was rightly pleased with the change, but the value stream manager knew much remained to be accomplished. She began using a version of the three-tier meeting process plus regular routine audits and gemba walks to generate task assignments to the supervisor, team leader, and support group representatives who worked with the area.

Just as you would like to see, the area did not rest on the accomplishments of the project team. Instead, it kept improving, driven by the ongoing process of assessments from the production tracking data, conversion into short-term assignments, and follow-up for daily accountability. This process went on for several months, becoming a routine. The value stream manager then initiated a suggestion system for process improvements. One of its features was that it was a visual system, described below. Second, and most important, was that it involved the supervisor and team leader sorting the suggestions and taking responsibility for getting them implemented in a few days or at most a week, in just the same way they had become used to taking responsibility for acting on daily improvement task assignments.

The value stream manager recognized that implementing some of the suggestions was beyond the scope of the team leader or even the supervisor. So, she separated the idea board into two segments. The upper half displayed the ideas and status of submissions from the team members that were being worked on or were in queue for the supervisor or team leader. The lower half of the display held team members' suggestions that the value stream's support groups were working on. The value stream manager held a portion of the stream's support group capacity in reserve for assignment to work on employees' worthy suggestions that were beyond the scope of local line management to complete. She held a weekly meeting with her value stream support group representatives, the area supervisor, and the team leaders to evaluate the week's ideas. At this meeting, they agreed on ideas to be assigned to support group members. Those idea cards were then moved to the to-do column in the lower half of the board with the assigned person's name noted on the card.

The result was a continuing steady flow of suggestions from operators who were reinforced by a steady stream of often modest improvements in the process, which continued to improve its performance. Not all the improvement was attributable to employee suggestions. Nevertheless, the team has remained open to change, in large part because they have the regular experience of being listened to when they make suggestions for change.

This is a powerful attribute, especially in the context of Lean conversion projects where the target is changing long-entrenched ways of doing things. Failure to listen to employees' ideas is often among those old ways. Coming from that kind of history, one might think frontline people actually have been "checking their brains at the door," and either do not have any ideas worth listening to or are uninterested in improvement.

In my experience, nothing could be farther from the truth. For one thing, the opportunity to be heard is powerfully motivating for people. This is true even when the only outcome is having been listened to. Further, in most cases, those on the frontline have not stopped having ideas. They have only given up on making suggestions. Indeed, in the project scenario above, when project teams ask for suggestions and feedback, they are typically inundated with ideas. The problem becomes the organization's inability to respond to them. At that point, ideas stop coming, and quickly.

Making Listening Visible

So, how can a visual control make listening visible and accountable and give frontline workers a sense of ownership and pride in improvement?

Use a format that encourages brief, readily displayed ideas. That is, require suggestions in writing on cards or Post-its®. This way, they are brief, easily displayed, and quickly moved. (For those who cannot write in English, dictating the idea to one who writes it down is perfectly acceptable.)

Keep your spreadsheet application in its holster; do not make lists of ideas! Remember the fingerprint factor. Keep and manage the ideas in the original form—the idea card. People have more sense of ownership when suggestions retain their fingerprints, their own handwriting and signature. Computer-generated lists can be intimidating, especially for those who do not work in this medium every day. If it is in the computer, the idea has become "yours." If it is in my handwriting, it stays "mine."

Create a visual representation of the way ideas move through the improvement process:

- First, ideas are submitted.
- Second, they are screened and either advanced to a queue or rejected.
- Third, they are actively worked on.
- Fourth, implementation is complete.

Such a process can look like Figure 11.4a and b. In it, you can see column headings for "ideas," "to do," "doing," and "done." As the cards move across

Idea Board: West Finishing Shop

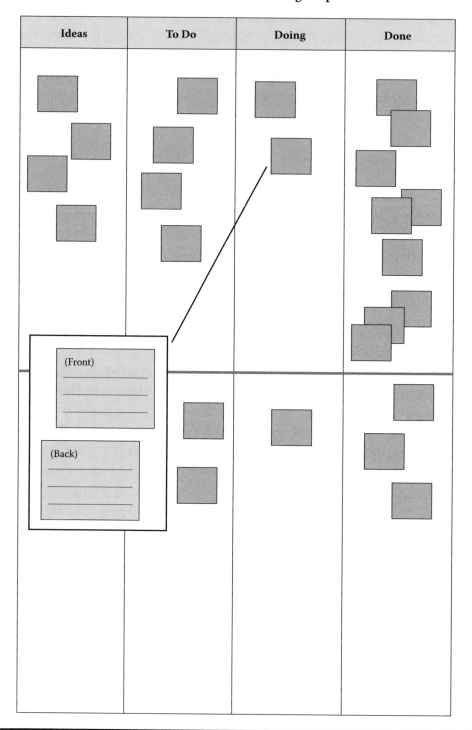

Figure 11.4 (a) Suggestion system idea board.

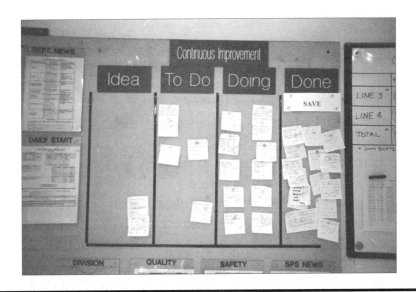

Figure 11.4 (*Continued*)   (b) Photo of an idea board.

the board and the tally of implemented suggestions climbs, it is difficult for holdout curmudgeons to maintain that "management never listens to us." Take them to the board, show the movement of cards, and point out the number implemented—each still in the handwriting of the person who made the suggestion—and suggest this way of being heard is open to them as well.

The process is uncomplicated, and it lends itself to variations that fit your particular circumstances and creative bent:

Step 1: Employees write idea cards (or Post-its), including their name, and post them in the "ideas" column.
Step 2: Once a week or more often, the supervisor and team leader review newly submitted ideas to advance them to the "to do" column or reject them. They note the reason for rejection on the back of the card, and talk with the author about the reason for rejection. In practice, few ideas are rejected. Reasons for rejection are typically scope (things for other departments to do), relevance to the business, or conflict with Lean principles.
Step 3: At least once a week, the value stream manager, supervisor, team leader, and value stream support group representatives review the new submissions. They identify ideas that are beyond the resources or ability of the supervisor and team leader. Those ideas get moved to the support group segment of the board (below the dividing line in Figure 11.4) and assigned to a specific individual.

Step 4: The supervisor and team leader move ideas from the "to do" to the "doing" column and assign them for implementation, noting the assignment on the card. The assignments are usually to the supervisor or team leader. The number of active suggestions is based on the capacity to get them done within a week. As work progresses or is completed, brief notes on the back of the card document plans and actions.

Step 5: As an idea is implemented, move the card to the "done" column and update the tally of implemented ideas.

Step 6: The team leader covers the status of the suggestion process once a week in a daily tier one start-up meeting, congratulating those whose ideas are done, or implemented, and reviewing ideas that have moved to the "doing" column.

Step 7: Some organizations use team rewards for reaching designated levels of implemented suggestions, such as pizza upon reaching one implemented suggestion per team member (or the equivalent number). Others find the motivational impact of people being able to influence their environment and being recognized for it is enough to keep the process healthy.

The improvement suggestion system is explicitly connected to the three-tier meeting process. The same expected versus actual accountability review applies to employee suggestions as with any other improvement assignment. Ideas assigned to team leaders show up on the department-level (tier two) task assignment board. Those assigned to support group representatives show up on the value stream-level task assignment board.

Quick Wins and Just Do It Processes

Some organizations effectively use a less structured approach instead of or as well as one like the idea board process, variously referred to as "quick wins," "just do it," or "quick kills." A worker or workers submit an idea and its anticipated benefits on a brief form. The supervisor or, more typically, the team leader okays it, and the worker or workers implement the suggestion during breaks in the schedule or in slower times. These programs typically occur in smaller work teams where team leaders are responsible for the day-to-day work process and are close to the hands-on or heads-down work. Descriptions or photos of these improvements are usually displayed once they have been implemented.

Lean Training for Line Leaders

A person's rank in an organization does not have much bearing on how he or she learns to be a Lean thinker and implementer. There is some benefit in learning from a book, class, or presentation, but people in any position really learn Lean by hands-on involvement and one-to-one coaching.

This is both useful and inconvenient! It implies that mass classroom training on Lean principles, though efficient, does not get the job done. Unfortunately, we cannot expect Lean training classes alone to transform first-line supervisors, team leaders, and frontline people into Lean thinkers.

It is not unusual for initial Lean training in an organization to come from outside, from a Lean consultant or sensei, as noted in Chapter 7. The sensei may do some limited classroom training but mainly the sensei spends one-on-one time with executives, plant managers, and perhaps with second-level production leaders, selected engineers, and specialists. One-on-one coaching from a sensei can be very effective, but it goes slowly and is costly. (Indeed, this is a constraint Toyota has experienced; its growth has outpaced its sensei resources.) As a result, the sensei's exposure to the rest of the organization is limited and, even with the original select group, typically does not continue beyond a year.

The key point is that hands-on experience and coaching is the way people learn Lean. This raises important questions: How can organizations bring personal coaching to the large number of team leaders and supervisors? And what are alternatives to the expensive external sensei?

Where Conventional Training Fits In

Conventional training does play a role in improving Lean knowledge and applications. My preference is for Lean training to be delivered by Lean implementers. After all, *sensei* translates as "teacher," or one who has gone before. The training delivery of an experienced Lean hand might not be polished compared to a professional trainer. But deficits in style are usually more than overcome by the authenticity that comes from personal experience, especially from within the organization. Many consulting firms and universities offer weeklong or longer Lean training programs. If you have no access to internal resources, these can be a starting point. But, keep in mind this Lean advice: smaller quantities more frequently, delivered to the point of use. As in Case Study 11.2, where you can, provide Lean training when and where it is needed on the tool, skill, or principle to be used. Where possible, avoid

CASE STUDY 11.2: TRAINING BY THE BOOK OR BY THE FLOOR?

A new manufacturing VP thought more Lean training was the answer to improving performance, and requested help from the corporate training group. A year later, corporate training delivered a curriculum that filled a 3-inch binder. The contents were organized in modules, each covering a separate topic such as value stream mapping, plan for every part, kanban pull systems, leading team start-up meetings, and standardized work. The modules were lengthy, each at least half a day in class, and included classroom tests and assessment forms to rate applications on the floor.

Some plants ignored the material; others struggled to deliver it. In the company's most advanced Lean plant, the plant manager and Lean leader knew they needed a better way to develop the Lean skills of team leaders and supervisors. They decided to use what they could from the binder.

Phase 1: Training. With help from their plant trainer, they revised and shortened each module to no more than two classroom hours. The plant manager told the plant's support group leaders, team leaders, and supervisors that to remain in those positions, he expected them to complete all the (revised) modules and pass each test. The plant devised a flexible sign-up schedule for classes, and also put the modules online. A few team leaders chose not to complete the series and returned to production jobs. The plant staff and the rest of the team leaders and supervisors completed the modules and passed the tests. So far, so good.

Phase 2: Assessing on the job. Several months after the training was completed, the plant manager and Lean leader recognized the Lean training had been only partially successful. Some of the "graduates" were effectively able to use the skills and tools in their daily work, while others were not. Again, the plant modified the material in the training binder to fit its needs. This time, it reworked the assessment for each training module into a development plan, expanding a generic numeric scale to descriptive comments on each module's practices, noted in three categories: below expectations, meets expectations, exceeds expectations.

Blank development plans in hand, they met individually with a small pilot group of team leaders in their own areas to assess the practices covered in the applicable training module. The pilot revealed different gaps in performance from a variety of causes. Some leaders understood a concept but not how to apply it. In other cases, a given module did not directly apply in a particular work area, such as flow design in a lot-by-lot fabrication area. Some lacked the computer skills to create or update forms or tracking charts, or needed coaching in leading meetings, or in interpersonal skills.

Phase 3: A team of coaches. The assessments and development gaps uncovered a need for tailored follow-up. The plant decided to experiment with one-on-one coaching using its own resources. It assembled a coaching team of individuals with the skills needed to address the development plans. One of the coaches had deep technical Lean expertise, one was good at spreadsheet programs, and another coached interpersonal skills and how to prepare for and lead stand-up meetings. An individual team leader might have 20- to 30-minute coaching sessions from one, two, or three coaches a week, on the floor in his or her work area.

Phase 4: Expanding development. The pilot leaders were reassessed after four to eight weeks of coaching. All had closed the gaps identified in their development plans. Based on this success, the plant extended the approach to all the team leaders, again a small number at a time to match the capacity of its coaching team. During this process, it became clear that some supervisors had development needs similar to those of the team leaders who reported to them, so the plant began including developmental assessments and coaching for supervisors as well, particularly on the skills needed to be more effective coaches for their team leaders.

The outcome: Overall, most of those who passed the training module tests were able to close the gaps identified in their individual development plans, but some did not. The plant manager, reflecting on the process and improved Lean skills among plant leaders. He calls the multiphase approach of training, assessment and development plan, and coaching a critical factor in the plant's continued Lean progress and improved operating results.

training that batches all the tools and principles together; in other words, avoid overproduction in training just as elsewhere.

Where you can, understand the need for a specific Lean tool or concept, and then focus training on it. Consider needs like these: the ability to do root cause problem solving to eliminate a flow interrupter, knowing how to use a machine balance chart to calculate kanban quantities, or learning to make observations in order to balance work so it flows among people in a work group. This is a more effective approach than referring people to tab 11 in the binder they brought home from their weeklong training session. When using this modular approach to training, you can increase its effectiveness by immediately assigning students to apply in their work areas what they have learned in class. Then, follow up by assessing the application and giving feedback on what was done well and what could be improved. The close linkage of concept and real-world application can be powerful.

Knowledge, Practice, Feedback: The Role of Coaching

Sometimes, that "ah-ha!" experience is enough for an individual to firmly cement the understanding and how-to skill, but not very often.

When the sensei works with a student, he or she tailors the approach to the individual. What part of a concept have you mastered? What needs more work? If you can see the need for an application in situation A, can you recognize the concept's application in situation B, in a different part of the operation, or in an altogether different part of the enterprise? A repeated algorithm behind the sensei's approach to teaching and learning is "knowledge, practice, feedback." The sensei should work with you, as long as you show motivation and progress, until you can see, for example, how the concept of load leveling or production smoothing (heijunka) applies in an engineering, healthcare, or marketing department just as it does in a physical production value stream. In fact, taking office and technical-professional people to a manufacturing setting can be helpful. Seeing in a physical, three-dimensional operation an otherwise abstract concept can help make the connection between concept and its potential applications.

Lean knowledge comes from practice in seeing and in doing. Knowledge, whether from training or another source, is only the beginning. Knowledge along with practice and feedback (an application of plan, do, check, act) leads to knowing *what* the concept is, *why* it is important, *how* it works, recognizing *where* it can apply, and being able to *implement* or *teach* it in disparate and apparently unrelated circumstances.

Conventional training, regardless of approach, can open the door to the Lean journey, but coaching is needed to see the path and progress along it.

What If Frontline People Don't Buy into Lean?

Problems with buy-in are almost always problems with leadership. These often include at least some of the following: a poorly articulated or weak case for change, failure to respect people's legitimate questions, not setting clear expectations at all levels, and weak or inconsistent follow-up on newly announced accountabilities and processes.

Even when none of these problems is present, some frontline people are just ornery, whether on the production floor, in the office, or in service positions. They bring a variety of personal and personality problems to work that lead them to refuse to accept the team leader's authority. That is especially a problem when you have just established team leader positions. A few people are likely to test the system in ways that can be difficult or impossible for a supervisor to observe or document.

Providing team leaders with a measure of authority is an effective way to respond to these initial challenges, as well as those that arise later from time to time. This stops short of including team leaders in the process of administering formal discipline. That should be left to supervision. Instead, it involves authorizing team leaders to make documented observations of problem behaviors that the supervisor can act on as a basis for disciplinary action.

That is not the same as the team leader administering formal discipline, and the documented observations do not always lead, and do not require, the supervisor to take disciplinary action. Further, each instance that results in a team leader's documented observation should be part of a conversation between the team leader and the employee in question. The authority comes from the fact that the team leader's notes are a sufficient basis, by themselves, for supervisors to take such disciplinary action as they see fit, without needing to have observed the behavior themselves. The effect on the responsiveness to team leaders' requests and suggestions is positive and dramatic.

Several conditions must be present for this process to be effective:

■ First, supervisors and team leaders need to reach a shared understanding of what constitutes enough to trigger documenting a problem.
■ Second, the supervisor must follow up on the team leader's action, if only with a conversation with the employee acknowledging the incident.

Table 11.1   Typical Items on Team Leader Notes

Thanks for	Please Work to Improve
Volunteering	Starting/stopping work on time
A positive attitude	Keeping up with standardized work
Offering a suggestion	Handling kanbans properly
Preventing a problem	Meeting requirements for quality
Extra effort	Following 5S standards/procedures
Other:	Other:
Team leader comments	Team leader comments

Otherwise, employees will have no more reason to pay attention to the team leader's requests than before.

■ Third, the process has to be simple and easy to use.

One example of this is supplying team leaders with a pocket-sized pad of preprinted notes. The notes list categories for behavior that needs improvement on the front side. On the back, to acknowledge and reinforce helpful behavior, the categories list positive contributions (Table 11.1). When the team leader observes either positive or problematic behavior worth noting, he or she talks with the employee, shows him or her the note, and then signs it and gives it to the supervisor. The supervisor responds within a shift, either talking with the team leader to better calibrate standards or talking with the employee.

Of course, it is important to be sure in advance that team leaders have the interpersonal skills to handle these kinds of interactions potentially involving conflict. It is important for supervisors and team leaders to reach a mutual understanding as to what kind of behavior warrants what kind of response, and timeliness of follow-up.

Responding to Low Performers

As work becomes balanced and flow depends on everyone in the system meeting expected outcomes, low performers show up like they are under spotlights. These low-performance situations can be troubling for leaders to deal with, but keep in mind that everyone in the workplace is watching what you do. Does everyone have to keep up, or are we willing to sacrifice performance for one or two? There is a direct 5-point checklist to review in determining what to focus on when working to turn around problems in individual performance.

- Are the tools and equipment the person is using calibrated and working properly?
- Are parts and materials they are using within specifications? Or, is the information they are using current, complete, and accurate?
- Has the person been appropriately trained?
- Have expectations for performance been made clear?
- Has there been regular feedback on performance?

If you rule out these benign explanations for a person's inability to do the work in a newly Lean area, your options become limited. As you go through these considerations, it can be helpful to keep a distinction in mind. There are those who can't do the work, perhaps unable to keep up in a takt-paced or high-volume setting. Then there are those who won't do the work, for many reasons. Your organization may have a place for people who can't meet expectations in a given production environment. Whether or not such a haven is available, you almost certainly have a progressive discipline system. You may be used to using discipline only in cases of objectionable conduct, especially in a production workforce. More likely you use the discipline system for performance problems among the salaried workforce. You will need to seriously consider using your progressive discipline system for performance in the factory, office, or service delivery setting as well.

Progressive Discipline

In these instances, the use of progressive formal discipline is an unambiguous sign to the employee that the performance problem is a real one that might eventually cost the employee his or her employment. When the alternative becomes unavoidable, some from the won't group suddenly become able to do the work everyone else does. In other cases, formal discipline is an increasingly clear signal for the person to find another position to move to where he or she can meet expectations, if such a position exists.

This is not a happy situation to encounter, but it is an implication of moving to well-defined and documented work processes, with clear expectations for outcomes overall as well as step-by-step. If these cases are not managed, many in the operation will find commitment to the Lean initiative open to question. It will be that much more difficult, if not impossible, to develop disciplined adherence to standards if the standards do not apply universally.

Human Resources Policy Issues in Lean Management

Lean management will almost inevitably involve changes to your organization's human resource policies. Any change is best accompanied by a restatement of the business case for the change to Lean production and an explicit connection between the case for change and the specific change at hand. Table 11.2 summarizes some of the policy areas that may be involved, their connection to Lean work processes, and some potential obstacles to overcome in making the change.

The changes in policies may be dramatic and far reaching, such as changing hourly pay systems from piecework to a flat or day rate. They may involve changing the policy that governs job elimination related to process improvement activity and subsequent exposure to layoff. They may involve the kinds of changes in authority and application of the discipline system outlined above. Some changes are more mundane, like changing break or start times, though any and all of these changes are capable of sparking emotional reactions. Having a firm grasp of why you are making the change and anticipating the questions and reactions you are likely to face are important preparation for working through these potentially contentious issues. Keep in mind that the best reaction often has nothing to do with stating the logic behind the change. Giving people the opportunity to make their displeasure heard is often the most effective thing you can do, especially since you are unlikely to be able to satisfy the desire to turn back the clock to the way things were before.

Involve HR in Lean

Your HR group is more likely to respond to your requests for support if they know something about the rationale for converting to Lean production. HR is likely to be interested in how Lean is changing the shop floor, office, and service delivery process, people's jobs and access to information, and their opportunities to participate in changes that affect them. Involve HR as much as you can, as early as you can—changing policies can take quite a bit of time in many organizations.

Take HR executives to the floor and show them what the new ways—and new performance measures—look like in comparison to the old. Share with them the case for change. Introduce them to the statistics on ergonomic benefits from the new job designs, process documentation, and rotation. Show them the visual proof of your methods for involving employees, for listening

Table 11.2   Potential Policy Issues in a Lean Conversion

Policy Area	Link to Lean Production	Potential Obstacles
Rotation	Rotation mitigates risk of repetitive stress injuries in work elements repeated at a takt pace; it results in a multiskilled workforce with many able to step in when needed; and it provides many eyes on each job, which increases the chances to see and suggest improvements.	Rotation must apply to all or it may be unenforceable. When initiated, not all may be able to succeed at each job in the rotation pattern. Will that disqualify those who cannot meet quality and takt requirements? What options will they have?
Layoff	Even though Lean will result in elimination of some work, nobody will lose employment as a result of process improvement. Lean should make us more competitive, preserving jobs in the long run. Layoffs might be needed if business conditions change.	Are you willing to temporarily absorb employees made surplus by Lean improvements? If not, forget about employees' cooperation and involvement in improvement.
Classifications and grades	Lean works best with a flexible, multiskilled workforce. Specialized knowledge is now contained in standardized work; previously complex jobs have been redesigned to support flow or to make them easier, like quicker setups. Existence of many grades and classifications is no longer warranted because of the changes in the jobs.	Many are proud of the grade or classification they have achieved and will see consolidation as a loss. Are you willing to work your way through this with your people? Can you reach agreement with your union, if applicable, balancing other changes with this one?
Pay	Lean works best with a flexible, multiskilled workforce. Because work has been restructured into smaller elements, and we have begun rotation and consolidated grades and classifications, the pay system needs to change to catch up with changes on the production floor.	Reducing distinctions in pay may end up reducing the pay of some employees. Are you willing to work your way through this with people, perhaps by phasing in the change?

Continued

Table 11.2 (*Continued*)   Potential Policy Issues in a Lean Conversion

Policy Area	Link to Lean Production	Potential Obstacles
Common or synchronized start times	Lean reduces buffers of inventory between processes. To maintain Leaner, lower levels of inventory, production needs to begin and end at specific times so we make what we need when we need it.	Start times can be surprisingly emotional. Are you willing to work your way through this with people? Can you phase in this change to give people time to adjust personal or family arrangements?
Common or synchronized break times	We need to make what we need when we need it. Synchronizing breaks in continuous process areas may be required for that. With balanced, takt-paced work, when one person leaves, everything stops. That means when one breaks, all break.	Some have been able to manage their own schedules, including longer or extra breaks. You will have to be willing to enforce break times more than you may have in the past, often an unpleasant duty.

and responding to their suggestions. Enroll them on your team; they'll come to help out eventually, if not right away.

Summary: Resolving People Issues to Support Lean Production and Lean Management

Predictability in the daily availability of people and a structured approach for responding to unplanned absences would be desirable in any production environment. It is especially important in a Lean production environment paced by takt time. Lean management provides a suite of labor planning tools that make attendance more visual, thus raising the level of public accountability for coming to work. Lean management's labor planning tools bring stability at least to the process of responding to unplanned absences.

An effective employee improvement suggestion system can be deceptively demanding if those in leadership positions are not prepared to respond to what is, in effect, new work being delegated to them in the form of improvement suggestions. The vacation paradox builds new capacity for acting on suggestions. Until that capacity is in place, it is best to hold off implementing an idea system. Actual versus expected and visual control applies in managing the suggestion system just as it does in most of Lean management. Employees appreciate being able to see the progress of their ideas, and leaders benefit from the increased trust in Lean that comes from the visual evidence of having listened to and acted on ideas from frontline workers.

Training is often part of HR's responsibilities. Lean training is most effective when the training group partners with operating units first to understand Lean and the role of coaching in Lean learning. With this foundation, the training group can help develop the materials and approaches the operating units need to develop, train, and effectively prepare people for Lean skills and thinking throughout their workforce.

A Lean implementation can raise questions about a number of HR policies. Lean requires much more precision in execution than a batch-and-queue system, or in environments where the work process had previously been left to the discretion or preference of individual employees. So, issues that interfere with disciplined adherence to Lean processes must be addressed quickly. Equipping team leaders to make authoritative observations of problem behaviors is one step. Preparations for applying progressive discipline to "can't" and "won't" performers is another. Beyond that, a series of policy changes may be called for, ranging from pay and consolidation of grades and classifications to required performance in job rotation, alterations in layoff policy, and changes in break and start times.

It is a good idea to involve the HR organization early and thoroughly in Lean. With a context in which to view the requested changes to support Lean production and Lean management, HR is much more likely to understand and work to accommodate your requests.

Study Questions

1. What are the implications, if any, for employees who miss an abnormally large number of workdays? For employees who virtually never miss workdays? Is there any discernable effect on overall morale?
2. Are jobs and tasks regularly shared here, or do people always do the same work in the same place? Do you think this has any effect on the ability to improve processes?
3. Are there standards for jobs that people need to meet, or are people allowed to just do the best they can? Does this affect morale one way or the other?

4. Do people have a regular, effective vehicle for suggesting changes to the work process, tools, task sequence, or other aspects of how the work is performed? Are suggested changes regularly tried out and put in place? Do people have the opportunity to work on implementing their suggestions? What is the overall effect on morale of whatever the current suggestion process may be?

5. What kind of Lean training is done here? Classroom, self-study, hands-on coaching, PDCA (plan, do, check, act) experiential learning? How effective has it been in supporting a robust Lean initiative?

6. Is progressive discipline used here for performance problems as well as for problematic conduct? What is the effect on the sense of equity in the workplace?

Chapter 12

Sustain What You Implement

The purpose of Lean management is to sustain a Lean production system. Without a Lean management system, Lean production implementations often falter, sometimes fail, and virtually never deliver up to their long-run promises. So, what sustains the Lean management system?

In a word, it's you.

As a leader in your Lean production environment, you are the force that can motivate and sustain Lean management in your unit. That applies no matter what your position, whether you are responsible for a team or department, a value stream or plant, a business unit, or the organization as a whole.

Expectations for processes and the ability to compare actual versus expected are the threads that connect the elements in Lean management. The person at the top of the unit, however defined, is in the position to set expectations and, most importantly, to follow up on them. Defining expectations and holding people accountable to them is the key to a successful Lean implementation. The higher in the organization this extends, the better the chances for success.

Making accountability easier to see and execute is the objective that underlies Lean management's ways of thinking, its tools and approaches. But, do not confuse tools and techniques with the indispensable ingredient: you as the chief accountability officer. Without you, no tools, no processes, no books can make your Lean implementation a healthy, growing, improving proposition. This chapter shows how leaders must sustain the Lean management system.

You Already Have a Management System!

Why is accountability such a big deal, such an important factor in sustaining Lean systems? The reason is written in your organization's history: you already have a management system. Actually, you as an individual might be developing a new one, but almost everyone working in a newly converted Lean operation still has the old one pretty firmly in place. Because of this, even with all the discipline and accountability you can muster, you should expect to see backsliding in just about every aspect of Lean you implement. Do not expect to like it, but be prepared for it.

Why does this happen if you have carefully implemented Lean management along with Lean production? Recall Smokey the Bear and the difference between breaking and extinguishing habits. You should not expect the new ways to stick just because people have adhered to them for a day or week. The old ways developed over time. They cannot be extinguished on demand. The old habits will reassert themselves in people who are very good, even you! Remember, nothing worth doing stays done forever without diligence, discipline, and hard work. Case Study 12.1 illustrates these points. This cautionary tale only serves to illustrate that nothing worth doing stays done by itself. As much as Lean management is about anything, it is about this.

> **CASE STUDY 12.1:   DON'T SOLVE
> PRODUCTION PROBLEMS BY HEROICS**
>
> A new product was replacing an older one on a 90-day schedule, so volume estimates for the new product were accurate. The leadership team for the product start-up had considerable experience designing and operating two Lean conversions, but only briefly. The processes they designed for this product incorporated much of what they had learned. They implemented a mixed-model, one-piece-flow assembly cell. A dedicated fabrication area replenished what assembly consumed, based on visual pull signals from fabrication's supermarket. As the preparation and debugging period ended, they seemed ready to take on the production volume from the outgoing product that was being replaced by their cosmetically identical, but internally superior, new model. Even so, as production shifted from the old to the new product with a sharp increase in daily production, all manner of material supply problems

broke loose. These problems had been lurking, but now were flushed out of hiding by the product cutover and sudden fivefold increase in volume that came with it.

The assembly supervisor was a veteran of two previous Lean conversions and had seen the kinds of glitches that come with a new operation. In this case, the glitches seemed to have gotten the better of him. On the occasion in question, about a month after the cutover, shortage of a particular purchased part threatened to shut down his assembly operation. This was on the heels of many frustrating days of start-and-stop operation caused by other shortages. The supervisor decided to keep assembly running, opting to hold the unfinished units in a finished goods staging area until the missing parts could be supplied and installed.

He was found the next day in the staging area down on his knees installing the missing parts himself. He explained he had continued to build units the day before "just to get the order to the customer faster." When pressed, he had to admit his old-style "heroic" response had in fact created more delay. Standard operating procedure in the response system (that he had been involved in developing), when encountering a stock-out, called for stopping the assembly operation and its 25 or so people and calling materials management. Depending on the situation, the assemblers were either to stand and wait or work on "job jar" items until a resolution was reached.

The needed parts had been in the building at the time, on the receiving dock. It is quite likely the materials manager, who would have been summoned by the response system, would have been able to locate and deliver the parts to the line after an annoying but relatively short delay. Production could have resumed and the units would have been on their way to the customer a day earlier at the cost of a half hour of overtime.

Bear in mind, this case involved a supervisor with a history of commitment to Lean. He understood and practiced the principles well and was considered a good Lean supervisor. Yet, when the wind blew hard enough, the embers of the old habits—"keep it running; we'll fix 'em later"—flamed back to life. He needed to be reminded that when "rocks" are exposed in the system, workarounds do not pave a path to success in Lean.

Further, a look at the production tracking chart in his area showed no entries documenting problems with material availability or delivery.

The supervisor and his team leaders were too busy fashioning work-arounds to tend their own process and document reasons for missed pitches. Without these data, the supervisor and value stream manager were unable to show what were their biggest problems. The result was more energy put into expediting, firefighting, and finger pointing.

Without data to systematically identify frequency and duration of interruptions, nobody, neither support groups nor floor leaders, could create a picture of what was happening and act to identify and address causes. All anyone could do was react to the last incident. The result was to replace stock-outs with expedited overstock. The area was soon swamped with unneeded material. All of this occurred in a value stream that regularly held three-tier meetings led by a value stream manager with experience using them effectively! But here, in the absence of data, the daily assignments were mainly for expediting.

After a few weeks of this kind of chaos, the value stream leader realized that one of the team leaders in the assembly cell had to "stay home" in the cell. This team leader's mission, no matter what, was to faithfully complete the entries on the performance tracking chart each pitch, actual units produced versus expected and the reasons for misses. With this information, the value stream leadership team was able to Pareto-chart interruptions and take systematic action to uncover and then resolve the causes. Within a few weeks, the clutter in the area diminished and production began to stabilize.

What Should You Do?

Stick to what you have just implemented. You have installed the engine, drive train, and controls of the Lean management system. Do not leave it in the garage, waiting for the weather to get nasty before learning how to "drive" your new management system. Proficiency in Lean management is like many other things; you get better with practice. It shows when you have to perform under pressure.

Consider the steps you have taken; you have defined expectations for performance and implemented tools to compare expected versus actual execution. These expectations are defined in day-to-day, operational terms in leader standard work. An important element in this standard work is to regularly reinforce visual controls. The visuals reflect adherence to or variation from processes, expectations, and standardized work at the value-adding task level. Then, the daily three-tier meetings and the cycle of assessment, assignment, and accountability will lead to temporary countermeasures while causes of variation are found and eliminated.

Rely on Leader Standard Work

The steps you need to take should be documented in your own standardized work. Follow it as you would follow a recipe for success. Require and reinforce others to follow their standard work as well. Reinforcement comes by briefly reviewing each subordinate's completed standard work document every day. Respond in a timely and appropriate way to requests subordinates have noted on their standard work forms. Respond as well to the other things you see on the forms. Is there a pattern of missing a particular element or elements? Does what you observe in the production area square with what you see in notes made on the standard work documents? When you treat your subordinates' completed daily standard work forms as living documents that prompt action, you powerfully reinforce standard work as "the way we do things around here—now." This is especially true when problems come up in your area.

Keep in mind the team leader who said her standard work allowed her to turn down requests to do others' jobs that would have distracted her from her duties. Also recall the supervisor who was able to steer his way out of chaos by returning to his standard work. With it, he regained control with much less time, effort, and frustration than in the few days following his vacation. Leader standard work is the Lean management tool of highest leverage. Look to it to sustain your Lean management and production implementations.

Maintain the Visual Controls

Where you have implemented visual controls, follow up to be sure they are being maintained. Verifying that visuals are current and the information on them is accurate and clear should be one of the key items on your standard work.

When problems arise in areas without visuals, quickly develop a tracking process and analysis appropriate to the problem. That is, when a reason for missed pitch is vague to the point where you have to ask the person who wrote it "What does this mean?" you have the opportunity to discuss the characteristics of well-written problem statements. Remember the axiom: "No action without data (even if action and data are concurrent), no data without action!" When reasons for misses are absent from the tracking forms or not clear enough to form the basis for taking the next step, treat these occurrences as teachable moments. You will have to explain several times how the data from visuals lead to assessment, assignment, and accountability for putting countermeasures in place and eliminating root causes. As people begin to see long-standing problems resolved, based on what appeared on their visual controls, most eventually will make the connection.

Without the information from well-executed visual controls, you will be driving your Lean management system in the dark without either map or headlights. You can get somewhere that way, but you will have the occasional crash, and when you arrive, you will have no idea how to get there again.

Visuals give you the information you need to choose the direction to apply your resources for root cause improvement. With visuals, you can dig out of problems; without them, you are likely simply to keep on digging the same holes over and over again.

Conduct Gemba Walks Regularly

If you have had the benefit of gemba walks with a sensei, remember that others may not have had the same opportunities to learn and apply, had their efforts reviewed and critiqued, and had the chance to apply again. The mantle of sensei now is on your shoulders, even if you continue to work with your own sensei, and even if you do not consider yourself an expert.

The learning model for Lean management especially, just as for Lean production, is the master-apprentice relationship. When you gemba walk others, you accomplish several things. You provide others with an opportunity for tailored, one-on-one learning. You demonstrate the importance of going to the place, looking at the process, and talking with the people as a key in assessing process performance.

For executives, your gemba walks focused on the health of the Lean management system are the backstop, the fail-safe for the Lean strategy. Your discerning assessment of the understanding and application of Lean management guarantees the integrity of execution of the Lean initiative as a whole, and cues you to intervene where needed to restore its integrity.

For line leaders, in a structured, scheduled way you reserve time to observe people and processes, draw inferences about steps you need to take, and refresh your ground-level view of how the process is operating to bring to your tier of the daily accountability process.

When you are in the production area, be alert for even small incidents that signify old thinking or habits. As a teacher, use what you see on the production floor to refresh and enliven basic Lean principles; it makes it easier for people to make their own connections between Lean concepts and what it means to apply them.

Expect to Repeat Lessons

You should expect to teach many lessons repeatedly, even with your best people. Recall the supervisor kneeling in the shipping area completing the assembly work he had shortchanged the day before. Have patience with people, but only so much. Try using the rule of three before considering next steps with individuals who seem slow either in understanding your lessons or in taking them to heart. That is, give people three chances (which might be on successive weeks' gemba walks on a given point, for example) before beginning to discuss performance.

As you give feedback, be most firm with those in the positions of most authority. When you convert to Lean, everyone's knowledge of what it means and how it works starts at zero. Just because some are in positions of greater responsibility does not mean they automatically know more about Lean production and Lean management than their subordinates. Indeed, unless they have invested time in their own learning, whether in hands-on implementation activity close to the front line or, for executives, gemba walking to master assessment and coaching of Lean management applications, they likely will know less about Lean management than those below them.

"On Board" versus Active Involvement

This is not about who is on board with the new program. This is about learning to think differently through active involvement, whether learning to implement Lean or, for executives, learning to assess the status of Lean management. The leaders in your organization represent your greatest leverage. Develop an involving, engaging approach for their learning, such as described in Chapter 8. This means teaching them to assess the adequacy of Lean management implementations. Develop "lesson plans" for gemba walks, such as the Lean Management Gemba Worksheets described in Chapter 8 and found in Appendix C. Help your executives into the role of student by providing a repeatable structure and process for gemba walks, complete with questions to answer and a test of their proficiency at the end of the walk. Close the gap between executives and your Lean initiative. Create the link that has been missing in so many organizations' otherwise technically sound Lean implementations.

Keep Yourself Honest

I often hire a guide the first time I go fishing in a new area. I find it helpful to fish and learn from one with knowledge and experience when in new territory. Expert, experienced consultation is an essential ingredient in the mix as you pursue your Lean conversion. Many times, you will find your sensei in the ranks of outside consultants, but not always.

Using a consultant in Lean is like fishing in another way, too. Folk wisdom holds that if you give me a fish, I can feed myself for a day. But, if you teach me to fish, I can feed myself for a lifetime. In other words, sustaining Lean production and, especially, sustaining Lean management largely have to be do-it-yourself propositions. Yes, you can call in your sensei periodically to assess your status, but in most cases you will find that he or she is telling you things you already know, but have somehow overlooked, allowed to slide, or missed in some way. Is there an alternative to paying for periodic reminders by your sensei (who in any case soon will likely make the suggestion to cease these visits)?

Even if you relied heavily on a consultant to design your Lean production system—a practice I do *not* recommend—it is a practical impossibility to rely on a consultant to sustain your new system. Responsible consultants understand their role mainly involves knowledge transfer to build Lean capabilities in their clients' contacts, and thus in their client organizations. This is critically different from creating dependency on the consultant's skills and knowledge, which mainly benefits the consultant at the expense of lasting benefit to the client.

Ultimately, A Do-It–Yourself Proposition

The same holds true regarding Lean implementation project teams. You may have needed a team for some of the more technical and analytic aspects of technical design and implementation, but you cannot rely on it to run your area for you.

Table 12.1   What Consultants Can and Cannot Do for You

Rely on Your Sensei to	Rely on Yourself to
Teach Lean principles, techniques, how to see the difference between batch and Lean	Implement the new Lean system and Lean management
Offer advice, critique, suggestions, and prods	Make decisions regarding how to proceed
Stretch and challenge your thinking	Make decisions regarding how to proceed
Provide "ah-ha!" insights	Make decisions regarding how to proceed Create and maintain disciplined adherence to the system
Stimulate you to take action	Continuously compare actual to expected
Review and critique your Lean management practices	Work directly with your subordinates regarding executing their expectations and extending the system in their areas

Consultants, internal or external, can talk with your leaders about expected versus actual outcomes until they are blue in the face, all to no avail. Unless the boss makes it clear by his or her behavior that something new is expected, subordinates will continue to conduct business pretty much as usual. The everyday discipline needed to sustain and extend the gains from Lean and Lean management is *always* domestic, never an import (Table 12.1).

Assess Your Lean Management System

It is a good idea periodically to assess the overall status of your Lean management implementation. This is an application of plan, do, check, act (PDCA) thinking embedded in Lean, an idea derived from the Lean principle of pursuit of perfection. An assessment using the measurement provided here or another one should do three things for you.

First, the dimensions and questions themselves should help to clarify what you are working toward, for yourself and for the rest of your organization. In this respect, they are standards, and are better referred to in that way. They are not to be attended to only when assessments roll around. They represent steps toward Lean goals, things to keep in mind routinely. Because the assessment dimensions represent the standards you are striving to achieve, the questions should be known to all—in advance of an assessment and on a regular basis. It is common in Lean organizations to feel that the more you achieve, the more that lies ahead to be done. You will most likely come to refine and redefine your understanding of what it means to achieve a standard, raising the level of expected performance as you go. This will reflect your growing understanding that Lean is an improvement system.

Second, an assessment should tell you where you stand relative to your standards and relative to your earlier status. It gives you a data point with which to evaluate the effectiveness of the work you have been doing to improve since the last measurement. This is not exactly profound, but without a standard of comparison, it is easy to believe that all is going pretty well, people understand what they are being asked to do and why, and we are making progress. Good—now prove it!

Third, the results of an assessment will help you identify where you need to focus efforts to improve. They may also lead you to conclude that you or your organization needs another dose of targeted help from the outside, perhaps from peers in a noncompeting industry, from a professional association, or from a teacher. Either way, the results should help clarify a next set of goals for shoring up a category where results have not been up to par with others, or for choosing one or a few areas to focus on for improvement in the next assessment.

Details of the Lean Management System Assessment

The Lean management standards cover process and behavior, and apply with slight variations in office (administrative, service, and technical-professional) settings, as well as in manufacturing. There are eight manufacturing dimensions and eight office dimensions. The standards define five levels of system status. This section lists the categories and levels in the standards. Both sets of standards are in the appendix and are available as free PDF downloads at www.dmannlean.com

Recall from Chapter 8 the reformatted gemba worksheets, designed for ease of use when learning to assess Lean management applications on the floor, rather than learning to become a Lean management implementer. These worksheets appear in Appendix C and are also available free at www.dmannlean.com. A downloadable version of the radar chart assessment report (see Figure 12.1 for an example) in Excel format is also available at the web address noted here.

Dimension	Actual	Maximum
Leader Standard Work	3	5
Visual Controls-Process	4	5
Visual Controls-Support	1.3	5
Standard Accountability	4	5
Value Stream Maps	2.7	5
Process Definition	2.3	5
Process Discipline	2.3	5
Process Improvement	2.7	5
Problem Solve	3	5

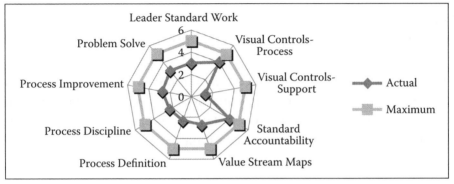

Figure 12.1 Score profile data and radar chart for Lean management assessment.

Lean management standards dimensions are:

1. Leader standard work
2. Visual controls—production
3. Visual controls—support (only in manufacturing standards)
4. Daily accountability process
5. Value stream mapping
6. Process definition
7. Disciplined adherence to process
8. Process improvement
9. Root cause problem solving

Levels in the Lean management standards are:

Level 1: Preimplementation
Level 2: Beginning implementation
Level 3: First recognizable state
Level 4: System stabilizing
Level 5: Sustainable system

Conducting an Assessment

The Lean management assessment should be conducted on the floor by looking and asking. Make direct observations (e.g., "Are the visuals current? Are the reasons for misses clear enough for taking the next step?"). Talk with people (e.g., "Do you have a regular way to make suggestions for improvement? Please tell me about it. In your opinion, how well does it work? Have you ever made a suggestion? Have people you work with made suggestions? Have any suggestions been adopted?")

Prepare a format that lists the criteria by level for each category to guide and record observations (or download the format based on one of those in the appendix). That way, documentation supporting a rating is right there with the criteria. Table 12.2 provides an example of such a format. Appendix A includes a complete Lean management assessment for manufacturing settings. Appendix B includes a complete Lean management assessment for administrative, service, technical-professional, and healthcare settings.

Table 12.2   Example of an Assessment Observation and Rating Form

Lean Management Standards: Manufacturing Leader Standard Work			Date and Location
Diagnostic Questions:			
1. Do leaders have standard work? Do they follow it? Do you carry it with you as a routine? Do you have it with you now?			
2. Is leader standard work regularly reviewed for updating? Has it been updated recently; when was the last time?			
3. Are standard work documents used as working documents, as a diary of the day? Are the leader standard work documents reviewed daily by leaders' supervisors? Weekly with the subordinate?			
4. Is there a regular place where completed standard work documents are stored? Is it visual? Is a leader standard work form available for all to see?			
5. Has leader standard work been used in this area in transitions between leaders?			
Progress			*Notes*
1	Pre-implementation	No leader standard work is in evidence.	
		Some leaders have heard of the idea of leader standard work.	
		Most leaders see standard work as applicable to production work only.	

Table 12.2 (Continued) Example of an Assessment Observation and Rating Form

Progress			Notes
2	Beginning implementation	Leader standard work exists for a few isolated positions.	
		Where it exists, leader standard work is carried some of the time and filled out irregularly.	
3	First recognizable state	Leader standard work has not been revised. Most leaders talk about it as a compliance, check the box exercise.	
		Standard work is in place for all leaders in the value stream, team leads through value stream manager. The standard work for all leaders has been revised once.	
		Most leaders carry their standard work with them, follow it, and use it as a working daily document.	
		Most leaders understand the benefits of standard work and can give examples of how it has helped them.	
4	System stabilizing	All leaders carry and follow leader standard work and use it as a working daily record.	
		All superiors regularly review subordinates' standard work documents daily.	
		All leaders can identify how standard work benefits them. Standard work is regularly reviewed and revised.	
5	Sustainable system	Standard work is regularly reviewed daily by the next level as a monitoring and communication vehicle. It is reviewed weekly jointly by subordinate and supervisor for patterns, revisions.	
		Completed standard work documents are visually maintained by day of week for each leader. A blank copy of leader standard work is posted in the area.	
		All transitions between leaders include review and walkthrough of leader standard work. All new leaders follow standard work from day one in the job.	

Few < 25%; Some < 50%; Most > 75%; All = 100%.

When Should You Assess?

Do it now. It does not matter how far along you are in your Lean journey. If you are just starting, an assessment will help you share expectations with others, and it will give you a baseline to use in the future so that you can judge progress and see how far you have come.

Of course, you have activities going on right now that will show improved results if you hold off on the next assessment so you can see the progress. But, that should always be true! Instead, establish an assessment schedule and stay with it. Measuring yourself every quarter is enough of an interval to expect to see progress—a long enough period of time in which to achieve improvement. It may also be a short enough interval to make it seem rushed, at least initially. An assessment is not a takt-paced activity, but assessments on a regular schedule embody the ideas of time and pace that are so important in Lean.

A related point is to keep the assessment process simple and free of bureaucracy. If you will be conducting an assessment every 90 days, then that is all the more reason you will want the process to be straightforward, rather than a burdensome time-eater. And, you will want the results in a timely fashion—the same or next day—so you can get started on the next set of improvements. After all, your next assessment is coming up in 90 days!

Who Should Assess?

You have several options. All of them should reflect the open-book nature of the assessment tool and process. That is, if the assessment categories and questions describe your standards for Lean management, would you not want them to be widely understood and practiced? One way to do that is just like the method you might use to calibrate with a subordinate on other assessments, such as a weekly 5S audit.

In the case of a 5S audit, you begin by reviewing the audit items with the subordinate (or group of subordinates). Then, give each an audit form and conduct the audit together. Compare ratings for each observation. Discuss points where you see things differently to develop a common understanding of what a particular item means and how it should be scored. The following week, conduct the audits separately and then compare results. Discuss points of difference, go together to look at the specific area where your ratings differed, come to agreement, and refine your understanding on how to score the item in question. Continue a second or third week until you are reasonably well calibrated. In this way, you have imparted your version of expected and how it compares with the commonly observed actuals.

You can use the same approach with the Lean management assessment. Work your way down through your organization, starting with the people who report to you directly. As practice, before you go "live," assess an area jointly with your direct reports or with your supervisor, comparing notes as you go. Then assess another area individually. Compare findings. Consider starting with a single assessment category applied across an individual's or staff's area of responsibility to get a good sample of observations to compare. When you are calibrated on that category, go to the next. This can easily be part of your regularly scheduled gemba walks.

As you complete the calibration process with this first level of your organization, ask them to continue the process with those who report directly to them. You might want to add an item to your standard work to spot-check calibration as the process spreads more widely, just as your standardized work calls for spot-checking execution of other items in your subordinates' standard work.

You will have to consider the size of your organization as you commence a regular program of assessment. Each unit down to the team leaders should assess its status every 90 days. Maintain a profile such as a bar chart or "radar" chart (see Figure 12.1) of category scores on display where you post your improvement plans for the period.

Wherever practical, a unit's assessment score profile should be based on the assessment by the leader of the next level in the organization—the unit leader's boss. At a certain scale, this will become impractical on a 90-day schedule. In those cases, consider a mixed model of assessors. Higher-level managers (plant managers, operations directors, VPs), where executives have learned to assess the status of Lean management, should participate in complete assessments of subunits on a regular schedule to maintain a common understanding of standards of assessment top to bottom throughout the organization. Regular assessments for large subunits, such as large value streams, can be conducted by a team with rotating membership drawn from the leaders of other units, along with a core made up of members of the organization's Lean team. These conditions will make it more important that assessment questions are reflected in the regular inspection items in the gemba walk protocols for leaders whose scope of responsibilities is large.

Interpreting the Assessment

It is more meaningful to look at assessment scores as a profile of the scores rather than as a single average value. That is because poor execution on one dimension can affect performance in other dimensions of Lean management, effects that spread through the others and eventually seriously compromise the way the Lean management system performs. An average could easily conceal this finding. And, averages provide no guidance on where to concentrate improvement efforts. Figure 12.1, a radar screen profile format, is an example of a directly interpretable profile format. Where it is meaningful to compare one unit with another or a single unit's performance over time, the total of category scores (with a possible range of 5 to 40), and a "consistency index" of the point spread between the lowest and highest category scores, serves as the best measure for overall comparisons. Again, an unaddressed low score in one category will eventually be reflected in low scores in others as well. Here, the consistency index will draw attention to gaps in performance.

Keep Asking These Questions!

Lean starts with physical changes. Until the "facts on the ground" or on the production floor are different, there is not much to be gained from implementing Lean management. But, as you begin making physical changes, do not implement them by themselves. Just as changes in the management system do not stand well by themselves, neither do technical changes. Every technical change requires support by the management system to maintain its integrity over time. Every time an element of Lean production is implemented, also put in place the elements of the management system needed to sustain it:

- A visual control tracking actual versus expected performance to capture misses and problems
- Elements in leader standard work to review the visual regularly and to lead an accountability process (if not already in place) to assign cause analysis and then action to eliminate the problems captured on the visual control

■ Elements in leader standard work to monitor for consistent execution of the newly improved process

As a check, each time a technical or physical element of Lean is put in place, including each time a change is made for improvement, ask these questions:

1. What Lean management practices must accompany this element or change to sustain it as it was intended or to test its effectiveness?
2. How will these Lean management practices be sustained?
3. How will you verify and monitor normal operation of the technical process?

A sensei trained in the Toyota approach to Lean production asks three questions about just about everything:

1. What is the process?
2. How can you tell it is working?
3. What are you doing to improve it (if it is working)?

The additional questions I am suggesting may be implicit in these Toyota questions. No matter, ask them explicitly! Table 12.3 illustrates how some of these questions can be put into effect.

A Lean Culture Is a Beautiful Thing

Perhaps the best way to tell you are in an operation with a Lean culture is to make a return visit to it. You should not have to be away long, only a few weeks. When you return, you should see things have changed. The changes do not have to be big; in fact, they can be pretty small in scale. Nevertheless, you should be able to hear people talk about the changes with a mixture of satisfaction and critique—like the do-it-yourselfer who talks with pride, yet points out the flaws in the beautiful piece of work he or she has just accomplished with his or her own hands. You should expect to hear something like this: "It is different than it was before. Here is how and why those differences are improvements. This is what remains to be done and what we are working on right now. Come back in a week and we will really have something to show you!"

Table 12.3   Some Answers to Lean Management's Questions

Element of Lean Production System	Element of Lean Management System
Pull system supermarket	Supermarket daily/weekly audit process; current findings and trend posted and coded green/red
	Visual display of "to be ordered," "ordered and due," and "overdue" deliveries
Kanban replenishment system	Comparison of actual replenishment cycle time versus standard setup plus run time for each order; reasons for misses noted
Flow line balanced to takt time	Hourly or more frequent production/pitch tracking versus goal and reasons for misses noted
	Daily value stream performance and task follow-up accountability meeting
Team leaders	Team leader standard work, supervisor and value stream manager standard work
Water spider line-side supply	Timed standardized route; comparison of actual route cycle time versus standard time for each cycle; reasons for misses noted; route standardized work regularly audited
Lean implementation activity of any kind	Daily or weekly gemba walks with a Lean teacher making and following up on assignments for improvement

The vision for the future can take shape in the present when people become used to thinking about making things even better than they are today. In this way especially, a Lean culture is a beautiful thing!

Summary: Maintaining Lean Management

Nothing sustains itself, certainly not Lean production or Lean management. So, act to sustain what you have done by following the processes you have implemented:

1. Use your standard work to establish or stick with a routine for monitoring your processes and the standard work of others. Remember, leader standard work gives you the most leverage in Lean management.
2. Check the status of visual controls as part of your routine. Insist that those responsible keep them current with accurate and complete entries

where reasons are called for. Teach people, more than once where necessary, why visual controls are important, where they fit in Lean management, how they drive action for improvement, and how that means both workers and leaders having a better day at work.

3. Follow up on what you expect in your daily three-tier accountability meetings. Assign tasks to stabilize, diagnose, and improve your area. Follow up on assignments and use visual accountability tools, for example, assignments posted on due dates and coded green for complete, and red for overdue. Do not shrink from green/red color coding.

4. For line leaders, schedule and stick with regular gemba walks with each subordinate. Stick faithfully to the schedule. On the walks, ask them questions to test whether they see what you see. Get a feel for the Lean concepts they recognize, grasp well, and can critique. Identify where their understanding needs strengthening. Give homework tasks to develop understanding, and follow up on assignments the next week. Take notes on your gemba walks; expect your students to do likewise. Remember, the first purpose of gemba walking is to teach.

5. For executives, develop mastery in assessing the quality of Lean management applications, as well as the understanding of Lean management among front line people and their leaders. Gemba walk on a regular schedule. Follow up when your assessment detects weak links somewhere in the chain of command.

6. Everywhere, ask: What is the process here? How could someone tell? Is it working? How is it sustained? Where necessary, note task assignments and gemba walk topics based on the quality of answers.

7. Establish an assessment schedule and a plan to phase it in. Share the detailed assessment categories widely. Post results where they can be seen. Expect to see evidence of improvement activities, such as A3s, daily task assignments, and other process improvement work to address low-performing categories.

8. Realize you will never be done and take steps to avoid burnout for yourself and others. Organize a regular process for sharing internal best practices so you and your team can recognize the successes you have achieved, even as you gird for work on further improvement.

If you have not started, choose one item from this summary list and begin with it. Practice it, add another, practice both, add a third. With every step, you will find yourself further along your own Lean journey.

Finally, remember what sustains the Lean management system, and ultimately, the Lean culture that grows from it. It is you and the example you set by your disciplined adherence to the system you have put in place!

Study Questions

1. Can you identify individual leaders in improvement initiatives who take responsibility for setting expectations, following up, and holding people accountable? What effect has this had, compared with initiatives lacking this kind of leadership responsibility?
2. Do leaders have standard work in your workplace? If so, what effect has it had on consistency and performance?
3. Are visual controls, if you have any, kept current in your workplace? Either way, what has been the effect?
4. Do leaders here regularly gemba walk in production areas? Are these inspection and to-do assignment tours only, or do they also include assessment, coaching, and teaching?
5. Does your organization rely on outsiders to maintain discipline, or is disciplined adherence to defined processes maintained internally? Why do you think one way or the other is best?
6. Do you regularly assess your progress in improvement initiatives? Either way, what's the effect?
7. If you have not started, which single practice do you think you'll choose to begin? Why?

Appendix A: Lean Management Standards—Manufacturing

Leader Standard Work

Diagnostic Questions:

1. Do leaders have standard work? Do they follow it? Do you carry it with you as a routine? Do you have it with you now?
2. Is leader standard work regularly reviewed for updating? Has it been updated recently; when was last time?
3. Are standard work documents used as working documents, as a diary of the day? Are the leader standard work documents reviewed daily by leaders' supervisors? Weekly with the subordinate?
4. Is there a regular place where completed standard work documents are stored? Is it visual? Is a leader standard work form available for all to see?
5. Has leader standard work been used in this area in transitions between leaders?

Progress			Notes
1	Pre-implementation	No leader standard work is in evidence.	
		Some leaders have heard of the idea of leaders standard work.	
		Most leaders see standard work as applicable to production jobs only.	
2	Beginning implementation	Leader standard work exists for a few isolated positions.	
		Where it exists, leader standard work is carried some of the time and filled out irregularly.	
		Leader standard work has not been revised. Most leaders talk about it as a compliance, check the box exercise.	
3	First recognizable state	Standard work is in place for all leaders in the value stream; team leads through value stream manager. The standard work for all leaders has been revised once.	
		Most leaders carry their standard work with them, follow it, and use it as a working daily document.	
		Most leaders understand the benefits of standard work and can give examples of how it has helped them.	
4	System stabilizing	All leaders carry and follow leader standard work and use it as a working daily record.	
		All superiors regularly review subordinates' standard work documents daily.	
		All leaders can identify how standard work benefits them. Standard work is regularly reviewed and revised.	
5	Sustainable system	Standard work is regularly reviewed daily by the next level as a monitoring and communication vehicle. It is reviewed weekly jointly by subordinate and supervisor for patterns, revisions.	
		Completed standard work documents are visually maintained by day of week for each leader. A blank copy of leader standard work is posted in the area.	
		All transitions between leaders include review and walk-through of leaders' standard work. All new leaders follow standard work from day one on the job.	

Value Stream Mapping

Diagnostic Questions:
1. Are there documented plans for improvement visible in each area—at least each department?
2. Are current process improvement activities visual in each area?
3. Are value stream maps used to identify and track process improvements?
4. Are value stream maps available for current and future state?
5. Do future state maps show planned kaizens, completion status of kaizens, and specific targets for improvement in measures of process performance?
6. Who prepares value stream maps for the area? How many of the area's leaders are proficient value stream mappers?

Progress			Notes
1	Pre-implementation	Value stream maps are not in use, or were used once but are out of date. They are not part of the area's approach to improvement.	
		Few people if any in the area are proficient at value stream mapping.	
2	Beginning implementation	Value stream maps can be seen posted in the area, may once have been used as an improvement planning tool, but they are out of date.	
		Some technical specialists (engineers, Lean leads) know how to map value streams. Most line manufacturing leaders do not.	
		Value stream maps, when present, show the current state only. Or, if future state maps are present, they are out of date.	
3	First recognizable state	Most areas have visible plans for improvement; many of these shown as current and future state value stream maps with kaizen bursts showing planned improvements and reflected on A3s.	
		Some of the future state maps show planned improvement in specific value stream process measures (such as lead time, %VA time, yield, productivity, uptime, changeover times).	
		In areas with current value stream maps, supervisors and superintendents are proficient value stream mappers and drew their own maps.	
4	System stabilizing	All value streams display current state and 90-day future state maps showing improvement goals (measures) and activities (kaizens).	
		All supervisors in the value stream are proficient value stream mappers, draw their own maps, use mapping to systematically understand opportunities large and small.	
		Completion status of kaizens is shown on the value stream maps, linked to A3 boards, and reflected in status of progress against 90 day goals.	
5	Sustainable system	Value stream maps are regularly used in the area's communications. Front line leaders teach value stream mapping.	
		All team leaders are proficient value stream mappers. All departments and teams use posted value stream maps to show their improvement plans.	
		Each area's performance (down to the team) is reflected in the current state measures summary on its value stream map (e.g., lead time, %VA time, yield, productivity, uptime, changeover).	

Visual Controls

Diagnostic Questions:

1. Are visual controls in evidence for production processes in flow <u>and</u> pull areas? Are they current?
2. Are reasons for misses or interruptions described clearly and specifically enough to decide what next steps to take?
3. Are visuals regularly reviewed and used to drive improvement? How? Are improvements stimulated by visuals limited to crisis situations, or are there many, often small, improvements driven by visuals?
4. Are visuals in regular use for out-cycle tasks like water spider/delivery routes, operator-based maintenance tasks, etc.?
5. Are visuals revised and changed as conditions change and issues either emerge or are resolved? Example?
6. Are visuals regularly signed off/initialed by leaders in the area?
7. Are visuals self-documented with "who does what here when" information at or on the visual controls themselves?

	Progress		Notes
1	Pre-implementation	There are no visual controls in evidence.	
2	Beginning implementation	Production tracking charts are posted in flow areas only.	
		Production tracking charts complete for few days or intervals of observation. They are filled in irregularly or with a focus on production numbers only. Reasons for misses are absent or too vague for action.	
		There is no or only irregular daily review of production visuals; response to information on the visuals is either absent or irregular. Visuals are a check the box activity.	
3	First recognizable state	Production tracking charts in flow areas are filled out regularly.	
		Most reasons for misses are specific and can serve as basis for deciding on next steps.	
		Visuals are reviewed daily, and most of the time drive specific action assignments on an identified major interrupter or problem.	
4	System stabilizing	All reasons for misses are clear and actionable.	
		Visuals are reviewed daily, and regularly lead to specific action assignments on small as well as large flow interrupters and other problems.	
		Tracking charts are in use at pacemaker, in flow and in pull areas. Visuals are added and discontinued as needs change.	
5	Sustainable system	Visuals are in regular use for out-cycle tasks throughout the value stream.	
		All production charts are initialed several times daily by department and value stream leaders, and occasionally by plant managers and executives.	
		Visuals are regularly analyzed to identify most frequent interrupters or problems, which then drive problem solving and improvement implementation.	

Visual Controls, Production Support

Diagnostic Questions:

1. Are visuals in evidence for nonproduction processes such as labor planning, 5S, etc.? Are they current?
2. Can leaders explain how actual versus expected performance can be visually controlled in nonproduction tasks?
3. Is review of nonproduction visuals included in the appropriate leaders' standard work?
4. Are improvements stimulated by visuals limited to crises or are there many improvements, small as well as large?
5. Are visuals regularly reviewed and used to drive improvement? How? Example?
6. What visuals are used to display and monitor TPM schedule and performance, SPC charts, daily calibration, quality checks, etc?

Progress			Notes
1	Pre-implementation	No visuals exist.	
		Monitoring and reports of production support activities, if any, are on IT systems, in rarely-referred to books, or in local, informal, undocumented systems.	
2	Beginning implementation	Some leaders understand it is possible and the reasons for applying expected-versus-actual visual tracking for non-production processes.	
		Initial visual controls are posted for a few processes, but completion and review of them is irregular.	
3	First recognizable state	Many visuals are in evidence, most are current.	
		Most visuals are monitored and reviewed regularly as defined in leader standard work.	
		Reviews of the visuals drive action on some major issues.	
4	System stabilizing	Charts and tracking processes are in place for all recurring production support activities.	
		Reviews regularly lead to actions on small as well as large issues.	
		TPM tasks, schedules, and assignments visually displayed using heijunka methods, and TPM performance is visually displayed either at the machine, at the TPM heijunka, or both.	
5	Sustainable system	There is immediate (one-day follow up) on lapses in maintaining production support visuals.	
		Tracking data from production support visuals are regularly analyzed for trends to identify opportunities for improvement.	
		All problems identified in production support visuals are followed up for root cause solutions.	

Daily Accountability Process

Diagnostic Questions:

1. Do regular meetings focus on the status of processes as well as on results? How often?
2. Do start-up meetings have clear purpose and agenda beyond today's production requirements/issues? Examples?
3. Do regular meetings result in task assignments to improve processes, and follow up past assignments currently due?
4. How are improvement assignments managed: visually, by spreadsheet or list, or not systematically?
5. Do visual control charts results in task assignments to address interruptions or problems?
6. How many leaders are familiar with basic project management techniques such as job breakdowns, and use them regularly? Examples?
7. How well integrated are support groups in value stream improvement activities? Examples?

	Progress		Notes
1	Pre-implementation	Daily plant, value stream meetings focus only on traditional production/shortage issues.	
2	Beginning implementation	Team start-up meetings are held sporadically at the floor, department, and value stream levels.	
		Team start-up meetings often lack a clear purpose. Meeting agendas mostly focus on the production numbers, schedules, and hours.	
		Team, dept., VS meetings regularly held but few tasks assigned or followed up at department, value stream meetings. Attendance inconsistent. Many task assignments incomplete, many moved from original due date.	
3	First recognizable state	Most team, dept., VS meetings regularly held. Most use task assignments; follow up occurs at department and value stream meetings. Attendance is consistent. Most assignments are completed, most stay on the original date.	
		Task assignments are mostly made in response to major disruptions. The work breakdown approach is used in making some task assignments.	
		Many leaders consistently use the green/red color coding convention.	
4	System stabilizing	Meeting agendas are regularly followed, attendance is faithful.	
		Review of prior day's visuals results in assignments on small as well as large items (some of which are converted to A3 projects). Task assignments made from many sources, not just visuals.	
		Green/red coding is a regular practice. Task notes stay on original due dates Many completed tasks reflected in positive trends in value stream's measures.	
5	Sustainable system	Accountability is routine; boards and green/red coding are used effectively for long and short term assignments.	
		All supervisors grasp, regularly use basic project management tools in determining task assignments.	
		Appropriate support groups routinely participate in value stream accountability meetings and are integrated into value stream improvement activities.	

Process Definition

Diagnostic Questions:

1. Are there documented definitions for all production and production support processes? Where is the documentation located?
2. Is the documentation current; does it match actual practice?
3. Is standard work available for production tasks? For how many levels of takt time? Is it posted?
4. Are operator balance charts available, for each takt, and posted in the areas they reflect?
5. Are definitions available, and posted, for tasks in the management process (e.g., who fills in charts, standard meeting agendas, etc.)?
6. Are job breakdown sheets used for process documentation? For training? Who maintains them? Are they current? Examples?

Progress			Notes
1	Pre-implementation	Process documentation is either in books or in IT systems.	
		Most process documentation is out of date and does not match actual practice.	
2	Beginning implementation	There is evidence of discussions in progress to replace obsolete definitions with Lean visuals for process controls, process tracking, work instructions, training documents.	
		Operator balance charts are present in a few flow areas, though few if any are current, and few if any represent multiple takts.	
		Standard work is posted in flow work stations for one takt time.	
3	First recognizable state	Repetitive processes are defined by standard work charts.	
		Standard work charts are for one level of takt in areas that operate at multiple takt times.	
		Most non-or infrequently repeated processes are documented; or visual documentation is present in most.	
4	System stabilizing	Areas that produce at multiple takt have operator balance charts and standardized work for each level of takt.	
		Definitions are in place for all production and management processes.	
		Process definitions are kept at the point of use or application and are up to date with actual practice.	
5	Sustainable system	Expected performance for all processes has been defined, documented.	
		Actual practice matches process documentation.	

Process Discipline

Diagnostic Questions:

1. Are defined processes regularly followed (e.g. 5S, punctuality, non-cyclical audits, PPE, etc.)?
2. Do crisis situations result in short cutting processes (e.g., production tracking, rotation, kanban triggers, etc.)?
3. Are manufacturing process and process safety audits carried out? Are support process audits carried out? By the leaders in the area or by outsiders?
4. When audits or tracking turn up noncompliance or misses, are problem-solving tools used?
5. To what degree does process focus lead to process improvement activity? Is there observable visual evidence?
6. How regularly do leaders conduct gemba walks to teach as well as to inspect? How many leaders do so?

	Progress		Notes
1	Pre-implementation	Leaders' attention is mostly on expectations for results.	
		Lack of consistent discipline is evident in production scheduling, 5S, punctuality, material control, and most other processes.	
2	Beginning implementation	Processes followed when things run smoothly, but abandoned when problems arise.	
		A few leaders can speak to the Lean rationale for process discipline and sticking with it.	
3	First recognizable state	Most leaders focus on obvious processes, e.g. standardized work, production tracking and pitch performance.	
		A few leaders focus on other processes such as TPM, 5S, pull systems, punctuality, labor planning.	
		Most areas are doing a good, clear, specific job of recording why interruptions or misses occurred.	
4	System stabilizing	Process focus includes non-cyclical areas like standard work in production control, water spiders' reasons for delays, visuals for scheduled maintenance as well as operator maintenance.	
		Routine audits on health of pull systems, actual versus expected changeover times, water spider routes, and other processes.	
		Most leaders using process tracking data to identify and act on improvement activities.	
5	Sustainable system	There are regular and frequent reviews of all production and related processes, including routine audits to maintain processes (i.e., 5S, pull systems, TPM, labor planning).	
		All process misses beyond production tracking produce task assignments for improvement.	
		Paretos of reasons for misses across all processes drive improvements.	

Process Improvement

Diagnostic Questions:

1. Who usually gets involved in process improvement: technical types, leaders, IT, other support groups, suppliers, workers, etc?
2. Who would most leaders say are the people most responsible for process improvement?
3. How are assignments made for process improvement tasks? Are the assignments and their status visually maintained?
4. How typical is it for improvement assignments to end up with actual improvements having been made?
5. Are kaizens a regular part of the improvement process in the area? For what kind of things? Who participates; who leads them?
6. Does improvement work focus mostly on big, technically-led projects or are small improvements also pursued?
7. Is there a regular way for employees to suggest improvements? What percentage of employees make suggestions? How many are implemented (few, some, most, all)?

Progress			Notes
1	Pre-implementation	Improvements are made by formal project teams, or in response to catastrophic failures.	
		IT, Finance, HR, other support groups lead improvement projects.	
2	Beginning implementation	Project teams make small improvements based on feedback during initial debugging.	
		Most leaders see improvement as responsibility of technical groups like IT, finance, HR.	
		Suggestion systems may be introduced but are not sustained.	
3	First recognizable state	Most leaders see process improvement as an area for their involvement. Some line leaders are actively involved in supporting improvement activities in their areas.	
		Most value stream managers and some other leaders are using daily accountability boards to drive improvements with green/red coding. Some tasks are completed on time, result in improvement. Some use A3 boards for larger-scale improvement projects.	
		Most leaders have participated in kaizens, a few have led kaizens. Few or none are qualified to facilitate kaizens.	
4	System stabilizing	Most leaders clearly see process improvement within their responsibility and can give examples of their involvement. All leaders have participated in kaizens; most now regularly lead kaizens.	
		Most leaders are effectively using daily task assignment boards, weekly A3 reviews as demonstrated by reviews of the process, the boards, and the completed tasks.	
		Some leaders experimenting w/employee process improvement suggestion systems.	
5	Sustainable system	Task assignments from regular stand-up meetings regularly result in small and large improvements.	
		Visual employee suggestion systems are established, sustained w/steady input of ideas, output of implemented improvements. Improvement plans and targets visibly displayed at value stream, department info centers.	
		Many leaders qualified kaizen facilitators. Plants have Lean resource teams to support local improvement activities and train employees in Lean through rotational assignments for interested individuals who meet qualifications.	

Problem Solving

Diagnostic Questions:

1. How often are workarounds used instead of investigating and resolving underlying causes of problems?
2. How often do leaders rely on data and analysis to attack a problem vs. gut feel, intuition, or impression?
3. To what degree do leaders expect changes will expose previously unseen problems that cannot be specifically anticipated, but proceed anyway?
4. How frequently do leaders ask why something happened vs. just asking what will we do to get back on track?
5. How frequently are leaders involved in leading problem-solving efforts?
6. How well understood and widely used are problem-solving tools such as five-whys, eight-step problem solving? Do leaders teach problem solving?
7. How frequently do leaders raise expectations for process performance in order to uncover the next level of process interruption or problem?

Progress			Notes
1	Pre-implementation	Problem solving only focused on workarounds, not on finding what caused problems.	
		Where cause analysis problem solving is used, it is limited to formal technical project teams.	
		Leaders can't describe a problem-solving process, or if they can, rarely if ever follow it.	
2	Beginning implementation	Leaders have begun using visuals to collect problem data but place little emphasis on pursuing cause analysis.	
		The most common response to problems is still to workaround and cover the cause with buffers of inventory, hours, etc.	
		Evidence of one or a few attempts at systematic problem solving.	
3	First recognizable state	Leaders beginning to ask why and pursue root causes for major problems.	
		Workarounds are recognized as such; evidence of problem solving methods used to understand and attack need for workarounds.	
		Uncovering production interrupters still viewed as troubling surprises.	
4	System stabilizing	Many leaders now asking why and pursuing root causes for problems small and large and beginning to use some form of structured problem solving, at least the five-whys.	
		Leaders expect to surface rocks with process changes and to resolve them at a root cause level.	
		Many leaders are seeking to improve their processes.	
5	Sustainable system	Leaders regularly expect cause analysis and pursuit of root causes for problems large and small.	
		Routine, systematic use of problem-solving tools to seek root cause solutions.	
		Process designs and measurements tightened up to uncover next level of problem: stated ultimate goal is to have perfect processes.	

Appendix B: Lean Management Standards—Technical-Professional, Administrative, and Service Delivery

Leader Standard Work

Diagnostic Questions:

1. Do leaders have standard work? Do they follow it? Do they routinely have it with them?
2. Is leader standard work regularly reviewed for updating based on new issues and changes?
3. Are standard work documents used as working documents to record notes and observations?
4. Are standard work documents reviewed at least weekly with leaders' supervisors?
5. Is there a defined place where completed standard work documents are stored? Is it used?
6. Has leader standard work been used in this area to facilitate transitions between leaders?

Progress			Notes
1	Pre-implementation	No leader standard work is in evidence.	
		Some leaders have heard of the idea of leaders standard work.	
		Most leaders see standard work as applicable only in areas with repetitive work.	
2	Beginning implementation	Leader standard work exists for a few isolated positions.	
		Where it exists, leader standard work is carried some of the time and filled out irregularly.	
		Leader standard work has not been revised. Most leaders talk about it as a compliance, check the box exercise.	
3	First recognizable state	Standard work in place for all leaders in dept. or VS, team lead manager/director. The standard work for all leaders has been revised once.	
		Most leaders carry their standard work with them, follow it, and use it as a working daily document.	
		Most leaders understand the benefits of standard work and can give examples that illustrate how it has helped them in their work.	
4	System stabilizing	All leaders carry and follow leader standard work and use it as a working daily record.	
		All superiors regularly review subordinates' standard work documents at least weekly.	
		All leaders can identify how standard work benefits them and the process. Standard work is regularly reviewed and revised.	
5	Sustainable system	Standard work is reviewed at least weekly by the next level as a monitoring and communication method. Subordinates regularly review it w/superiors to uncover patterns and consider revisions.	
		A standard process is followed for turning in, storing, and reviewing completed standard work documents for each leader.	
		All transitions between leaders include review and walkthrough of leader standard work. All new leaders follow standard work from day one in the job.	

Value Stream Mapping

Diagnostic Questions:

1. Are there documented plans for improvements in each area in the value stream, or at least in each department?
2. Are process improvements posted and visual in each area?
3. Are value stream maps used to identify and track process improvements?
4. Are value stream maps displayed for current and future states?
5. Do future state maps show planned kaizens, completion status of kaizens, and specific targets for improvements in process measures?
6. Who prepares the value stream maps for the area? How many of the area's leaders are proficient value stream mappers?

Progress			Notes
1	Pre-implementation	Value stream maps are not in use, or were used once but are out of date. They are not part of the area's approach to improvement.	
		Few people in the area are proficient at value stream mapping.	
2	Beginning implementation	Value stream maps can be seen posted in the area, and may once have been used as an improvement planning tool but are now out of date.	
		A few technical specialists in the area know how to map value streams. Most of the leaders do not.	
		Value stream maps, when present, show the current state only. Or, if future state maps are present, they are out of date.	
3	First recognizable state	Most areas have plans for improvement posted, with many of these shown as value stream maps of current and future states with kaizen bursts indicating planned improvements also reflected on A3s.	
		Some future state maps show planned improvement in specific process measures such as cycle time, %C+A, process time as % of cycle time.	
		In areas with current value stream maps, some team leaders are proficient mappers and draw their own maps.	
4	System stabilizing	All value streams display current state and 90-day future state maps showing improvement goals for process measures and kaizens to achieve the goals.	
		Team leaders in the area are proficient mappers and draw their own maps. Mapping is a tool used systematically to understand opportunities large and small.	
		Completion of kaizens is shown on the value stream maps, linked to A3 displays, and reflected in status of progress against 90 day goals.	
5	Sustainable system	Value stream maps are regularly used in the area's communication process. Front line leaders teach value stream mapping.	
		All leaders (managers, directors, administrators, VPs) are proficient mappers and interpreters. All departments and team areas use posted value stream maps to show their improvement goals.	
		Each area's process performance (down to the work team) is reflected in the current state process measures summary on its value stream map (e.g., cycle time, process time as % of cycle time, %C+A, productivity, etc.).	

Visual Controls

Diagnostic Questions:

1. Are visual controls in evidence for balancing workloads To show pace and progression of work? For exception management?
2. Are reasons for misses and interruptions described clearly and specifically enough to decide next steps to take?
3. Are visuals regularly reviewed and used to drive improvement? Are the improvements stimulated by visuals limited to crisis situations, or are the many, often small improvements driven by visuals?
4. Are visuals regularly used for support tasks, e.g. materials, rotation, planned absences, skill versatility, ancillary service performance?
5. Are visuals revised and changed as conditions change and issues either emerge or are resolved?
6. Are visuals regularly signed off/initialed by leaders in the area?
7. Are visuals self documented with "who does what here when" information at or on the visual controls themselves?

Progress			Notes
1	Pre-implementation	There are no visual controls in evidence.	
2	Beginning implementation	Production or exception visuals are used only to show the status of work; who's doing what, and to balance workloads.	
		Tracking charts for value adding and support processes complete for few days or intervals of observation; often filled in irregularly, focus only on " the numbers." Reasons for misses absent or too vague for action.	
		There is no or only irregular daily review of process, service, or production exception visuals. Response to information on the visuals is either absent or irregular. Visuals are a check the box activity.	
3	First recognizable state	Some support processes and systems have visual tracking for completion or reliability. Tracking charts, where they exist, are current.	
		Most reasons for misses are specific, complete and lead to next steps with little additional information needed.	
		Visuals are reviewed daily or on a regular schedule, often drive specific action assignments on identified major interrupters or problems.	
4	System stabilizing	Tracking charts in use for each flow path, handoff, production, service area. Visuals used to balance loads and reflect expected vs actual pace or progression of work. Visuals added, discontinued as needs change.	
		All reasons for misses/exceptions specific, complete, lead to next steps with little added information needed.	
		Visuals are reviewed daily, regularly leading to specific assignments on small and large interrupters and other problems.	
5	Sustainable system	Visuals are in regular use for support, ancillary areas, and systems throughout the value stream.	
		All process, service, and production tracking charts are initialed at least daily by department and value stream leaders and occasionally by executives gemba walking in the area.	
		Visuals are regularly analyzed to identify most frequent interrupters or problems, which then drive root cause problem solving and improvement.	

Daily Accountability Process

Diagnostic Questions:

1. Do regularly scheduled meetings focus on the status of processes as well as on results? How often?
2. Do start-up or stand-up meetings have clear purpose and agenda beyond today's requirements?
3. Do regular meetings result in task assignments to improve service or processes, and follow up on overdue assignments?
4. How are improvement assignments and projects managed; visually, by spreadsheet or list, not systematically?
5. Do visual control charts result in task assignments, problem solving, or project assignments to address interruptions?
6. How many leaders are familiar with and able to apply basic project management techniques like work breakdown structures?
7. How well integrated are support, ancillary, or supplier groups in process improvement activities?

Progress			Notes
1	Pre-implementation	There are no regularly scheduled meetings to make or follow up on task assignments or projects for improvement or remediation.	
		Daily department and value stream meetings focus only on traditional production and volume issues.	
2	Beginning implementation	Start-up or stand-up meetings are held sporadically at the team, department and value stream levels.	
		Start-up or stand-up meetings often lack a clear purpose. Meeting agendas mostly focus on the day's goals and schedules.	
		Team, dept., and VS meetings regularly held, but little use of task assignments or follow up. Attendance is inconsistent. Many dates slip, no or inconsistent use of visual tracking for issues, assignments, or projects.	
3	First recognizable state	Team, dept, VS meetings regularly held using task assignments. Follow up occurs at department and value stream meetings. Attendance is consistent. Most task assignments are completed; most stay on the original due date.	
		Tasks mostly assigned in response to major disruptions. Some leaders use implicit work breakdown approach in assigning some tasks.	
		Many leaders consistently use the green/red color coding convention to indicate completed on time or overdue status of assignments.	
4	System stabilizing	Meeting agendas are regularly followed; attendance is faithful.	
		Review of prior day's visuals results in assignments on small and large items (some become A3 projects). Tasks assigned from many sources, including employee suggestions, gemba walks, not just visual tracking. Most leaders are familiar with and use project management logic for task assignments.	
		Green/red coding is a regular practice. Accountability tasks stay on original due dates. Many completed tasks reflected in positive trend in value stream, unit or department measures.	
5	Sustainable system	Accountability is routine in tasks and projects. Boards and green/red coding used effectively for long and short term assignments.	
		All leaders grasp, regularly use basic project management tools to determine task assignments, dependencies, and durations.	
		Supplier and customers groups routinely participate in accountability review meetings, are integrated into value stream, department, unit improvement activities.	

Process Definition

Diagnostic Questions:

1. Are there documented definitions for all production and support processes? Where is the documentation located?
2. Does the documentation match actual practice?
3. Are production, service, and process tasks and procedures documented in Job Instruction Training (JIT) format with Job Breakdown Sheets (JBS) and standardized work or procedures?
4. Are standardized procedures documented, available for recurring tasks and situations? Is this information readily accessible?
5. Are definitions available and accessible for management process tasks, e.g., who fills in charts, standard agendas, etc.?
6. Is documentation updated as processes change? Is this an assigned and executed responsibility? By whom?

Progress			Notes
1	Pre-implementation	Process definition is either in books or in IT systems.	
		Most process documentation is out of date and does not match actual practices, or is only available at the level of policy rather than specific task(s).	
2	Beginning implementation	There is evidence of discussions in progress to replace obsolete definitions with JIT/JBS and standardized work documentation, visual process or service progression tracking and/or production controls.	
		Standardized procedures are documented and available for some tasks and situations.	
		Work balancing/leveling and volume-dependent staffing procedures are available in a few areas or for a few tasks.	
3	First recognizable state	Standardized procedures are documented and available for most high volume/everyday jobs.	
		Job Breakdown Sheets and standardized work sheets are in use to document work instructions. Job Instruction Training is in use for training in most high volume/everyday jobs.	
		Standardized procedures and aids (checklists, templates, etc.) are available for high volume/everyday tasks.	
4	System stabilizing	Areas that produce at variable volumes/rates of demand have defined, documented roles, procedures, JBS training for variable staffing levels.	
		Job Breakdown Sheets and standardized work are in place for all production, service, ancillary, support and management processes. Responsibility for maintaining documentation is defined.	
		Process definitions are kept at or are accessible at the point of use, application, or service delivery and are up to date with current actual practice.	
5	Sustainable system	Expected performance for all processes has been defined and documented, often in Job Instruction Training forms and formats.	
		Actual practice matches process documentation.	
		Maintaining process definitions and documentation is a clearly defined, assigned, and consistently executed responsibility.	

Process Discipline

Diagnostic Questions:

1. Are defined processes regularly followed, e.g., 5S, standardized procedures, checklists/templates, regular meetings, etc.?
2. Do crisis situations result in shortcutting processes, e.g., process tracking, standardized procedures, etc.?
3. Are work processes tracked for compliance with standardized procedures and practices? By people and leaders in the area or by outsiders?
4. When audits or tracking turn up noncompliance or misses, are problem solving tools used to understand the causes?
5. To what degree does focus on the defined processes lead to process improvement activity? Is there observable evidence?
6. How regularly do leaders conduct gemba walks to teach as well as to inspect? How many leaders do so?

Progress			Notes
1	Pre-implementation	Leaders' attention is mostly on expectations for results. Consistency of practice and discipline to defined processes is generally lacking.	
2	Beginning implementation	Processes are followed when things run smoothly, and are abandoned when problems arise.	
		A few leaders can speak to the Lean rationale for process discipline and sticking with it.	
3	First recognizable state	Most leaders focus on obvious processes, e.g., standardized procedures for recurring tasks, production, service, and process tracking, measures of pace and timeliness.	
		A few leaders focus on other processes such as work balancing, cross training, controlled release of work, wait times, rooming and discharge processes, etc.	
		Most areas are doing a good, clear, specific job of recording when interruptions, delays, or misses occur.	
4	System stabilizing	Process focus includes customer and supplier areas, service delivery, ancillary and support activities such as system downtime, materials and supplies, cross training, wait times.	
		Routine audits take place to assess cross training, work balancing, proficiency with tools and practices, accessibility and up to date status of documentation for standardized processes and procedures.	
		Most leaders are using process and service tracking data to identify and act on improvement opportunities.	
5	Sustainable system	Regularly scheduled and performed reviews of production, service, and related processes result in appropriate, periodic revisions and updates of standards.	
		All process misses beyond production, service and related process tracking are considered for task assignments for cause analysis and/or improvement.	
		Pareto analysis of most frequent and/or serious misses across all processes drives short term improvement assignments and longer cycle (A3) improvement projects when warranted.	

Process Improvement

Diagnostic Questions:

1. Who usually gets involved in process improvement in this area? Specialists, leaders, IT, area employees, etc?
2. Do leaders teach problem solving?
3. Is there an organized, structured approach to problem solving? Examples, visible evidence for it?
4. Who would leaders say are the people with responsibility for process improvement?
5. How are assignments made for process improvement tasks? Are the assignments and their status visually maintained?
6. How typical is it for improvement assignments to end up with <u>actual</u> improvements having been made?
7. Are kaizens a regular part of the improvement process here? Who gets involved? Who leads the kaizens?
8. Does improvement work focus mostly on big, technically-led projects or are small improvements also pursued?
9. Is there a regular way for employees to suggest improvements? What percentage of employees make suggestions? How many are implemented (few, some, most, all)?

Progress			Notes
1	Pre-implementation	Improvements are made by formal project teams, or in response to major failures.	
		IT, finance, HR, quality, other support groups lead improvement projects.	
2	Beginning implementation	Project implementation teams make small improvements based on feedback during initial debugging.	
		Most leaders see improvement as responsibility of technical groups such as IT, finance, HR, quality.	
		First attempts made to introduce systematic problem solving and related leader training. Suggestion systems may be introduced but are not sustained.	
3	First recognizable state	Most leaders see process improvement as an area for their involvement. Some leaders actively support improvement activities in their areas. Some leaders teach problem solving. Evidence of structured problem solving.	
		Most leaders in area use accountability task boards to drive improvements with green/red coding. Some tasks completed on time, result in improvement. Some A3 plan formats in use for larger improvements, some use of future state VSMs to display current improvement plans.	
		Most leaders have participated in kaizens, a few have led kaizens. Few or none are qualified to facilitate kaizens.	
4	System stabilizing	Most leaders see process improvement as their responsibility, can give examples of their involvement. All leaders have participated in kaizens; most now regularly lead kaizens. All leaders teach problem solving.	
		Some leaders are experimenting with employee suggestion systems for process improvement. Systematic problem solving is in use by most and evident in the area.	
		Most leaders effectively use task assignment boards, weekly A3 reviews w/visible evidence of boards, reviews, improvements. Most use future state value stream maps to display planned improvements and goals.	
5	Sustainable system	Task assignments from regular stand up meetings regularly result in small and large improvements.	
		Employee suggestion systems established, sustained w/steady input of ideas, output of implemented improvements. Improvement plans, targets visibly displayed in the area. Problem solving activity and thinking in evidence among all in the area.	
		Many leaders qualified kaizen facilitators. Departments have kaizen/Lean resources to support local improvement activities, train employees in Lean through rotating assignments for those interested and who meet selection criteria.	

Root Cause Problem Solving

Diagnostic Questions:

1. How often are workarounds used instead of investigating and resolving underlying causes of problems?
2. How often do leaders rely on data and analysis to attack a problem vs. gut feel, intuition, or impression?
3. To what degree do leaders proceed with changes even though they expect that changes will expose previously unseen problems that cannot be specifically anticipated?
4. How frequently do leaders ask why something happened vs. just asking what will we do to get back on track?
5. How many leaders teach and lead problem solving efforts?
6. How well understood and widely used are structured problem-solving tools such as five-whys, eight-step problem solving?
7. How frequently do leaders raise their expectations for process performance and tighten process measures in order to uncover the next level of process interruption or problem?
8. In conversation, do employees demonstrate problem-solving, continuous improvement thinking?

Progress			Notes
1	Pre-implementation	Problem solving is only focused on workarounds, not on finding causes of problems. Most employees do not believe problem solving is part of their jobs.	
		Where cause analysis problem solving is used, it is by formally chartered technical teams or designated specialists.	
		Leaders can't describe a problem-solving process, or if they can it's not documented.	
2	Beginning implementation	Leaders have begun using visuals to collect problem data but place little emphasis on pursuing cause analysis.	
		The most common response to problems is still to workaround and/or cover the cause with buffers of overtime, extended lead times, temp help, etc.	
		Evidence of one or a few attempts at systematic problem solving. Some leaders trained to teach problem solving.	
3	First recognizable state	Leaders beginning to ask why and pursue root causes for major problems. Many leaders teaching structured problem solving, many team members getting involved in problem-solving.	
		Workarounds are recognized as such; evidence of problem- solving methods in use to attack them.	
		Interrupters uncovered after process changes still viewed as troubling surprises.	
4	System stabilizing	Many leaders now asking why and pursuing root causes for problems large and small and beginning to use some form of structured problem solving, at least the five-whys.	
		Leaders expect to surface " rocks" with process changes and to resolve them at a root cause level.	
		Many leaders seek to improve processes or services by using systematic cause analysis and problem-solving methods. Many employees involved in structured problem solving.	
5	Sustainable system	Leaders regularly expect cause analysis and pursuit of root cause solutions for problems large and small.	
		Problem-solving tools used routinely to seek root cause solutions. All leaders teach problem solving, all employees involved in problem solving, all in the area involved in problem-solving thinking.	
		Process designs and measurements tightened up to uncover next level of problem to attack: stated ultimate goal in the area is to have perfect processes.	

Appendix C: Lean Management System Gemba Worksheets

Visual Controls

Intent: Visual Controls should do at least one of two things:
- Reflect the actual vs. expected pace or progression of work (admin, support, or line processes)
- Capture delays, interruptions, and frustrations that arise doing the work

Diagnostic questions:

1. Can you see visual cycle or procedure tracking charts in the area? Do they show expected vs. actual times?

2. Are the charts current to this or last shift?

3. Are incidents that delay work described clearly (What we had but did not want, wanted but did not have)?

4. Are visuals reviewed regularly? How frequently? How can you tell?

5. Can leaders & task-level people in the area cite improvements from problems noted on visual charts?

6. Are visuals used here for support tasks, e.g., materials, transport, attendance, assignments, qualifications?

7. Do leaders regularly review the visuals? How often? How can you tell?

Assessment: Rate this area/areas from 1 to 5 on the scale below and note rationale for the rating

1. Pre-Lean	2. Starting	3. Recognizable	4. Stabilizing	5. Sustainable
No visuals/cycle tracking in place.	Some cycle tracking charts; irregularly filled in. Most charts record numbers, do not document delays, problems. Where problems described, too vague for action. No or irregular review for action on problems. Visuals more "check the box" than tool to highlight problems, delays, and drive improvement.	Many front line & support areas here use visuals/cycle tracking charts. Charts are current. Most descriptions of problems are complete, specific enough for next steps (cause analysis or corrective action). Charts reviewed daily or on regular schedule. Problems noted on charts often result in assignments for action.	Visuals used for most line, support, & admin activities here. Visuals used at most handoffs between functions/departments w/ regular joint review for action. Charts revised, added, dropped as things change. Nearly all problem descriptions clear, complete, actionable. Daily/regular reviews of charts drive assignments for cause analysis or corrective action.	Visuals/cycle tracking charts regularly used throughout the area, front line, support, and administrative activities. Visuals/tracking charts initialed at least daily by line leaders and occasionally by executives. Visuals/cycle tracking charts regularly drive improvements, are also periodically analyzed to identify and act on recurring problems.
Rationale for this rating:				

Standard Accountability Processes

Intent: Standard Accountability processes
- Accountability processes should convert problems/opportunities noted on visuals, the floor, or from suggestions to task assignments - for cause analysis and/or corrective action in a daily Post-It (or equivalent) process for briefer tasks, a weekly A3 process for longer ones.

Diagnostic questions:
1. How are improvement assignments and projects managed here; visually, by spreadsheet or list, or not at all?
2. Are regular (daily or weekly) meetings held here to make new task assignments to address problems and follow up on overdue assignments?
3. Do the regular meetings here have clear purpose and agenda - other than today's anticipated work? What is it?
4. Do visual controls/cycle tracking charts result in task assignments to address interruptions, delays, capacity losses?
5. How many area leaders are familiar with and able to apply basic project management approaches - like work breakdown structures and dependencies - in thinking through and defining task assignments?
6. How well integrated are support, customer, or supplier groups in this area's improvement activities?

Assessment: Rate this area/areas from 1 to 5 on the scale below and note rationale for the rating

1. Pre-Lean	2. Starting	3. Recognizable	4. Stabilizing	5. Sustainable
No regularly occurring visual process to make or follow up on task assignments for improvement based on identified problems or delays.	Daily or weekly start up/team meetings held regularly for improvement task assignment; many are completed on time. Many assignments are to support or admin groups vs. line area, or are made in response to major problems. Many using green/red coding for on time completion or past due tasks.	Team or area (line & support group) meetings regularly (daily/weekly) held to make, follow up on improvement task assignments. Tasks posted visible to all. Attendance is consistent; most tasks are completed, most on time, most leaders use green/red coding for on time or late task completion. Tasks respond to both major, minor incidents. Much reference to customer/user/patient perspective.	Accountability meetings crisp, agenda followed, attendance faithful. Small assignments to visual accountability board; larger ones to A3 projects Green/red coding is routine. Tasks from many sources, not just visuals but also employee suggestions, gemba walks, support areas. Many in area use project management skills on project work. Customer perspective is a given.	Using the accountability processes is routine in the area. All leaders regularly use basic project management tools to determine task assignments, dependencies, durations. Support and admin representatives routinely participate in line accountability process and have their own. Customer's perspective informs most assignments, admin, support, frontline.

Rationale for this rating:

Leader Standard Work

Intent: Leader Standard work should reflect process focus:

- The closer to the task execution level, the more frequent the focus (admin, support processes, production/pt. care)
- Should reflect "go to the place, talk with the people, observe the process" for all levels of leadership
- Review of visuals (current? quality of entries? regular in-shift review?), accountability (assignments linked to problems from visuals), follow up on improvements in leaders' std work (faithful execution of redefined processes)

Diagnostic questions:

1. Do leaders in this area have standard work? Do they follow it? Do they routinely have it with them? Can leaders describe how standard work has helped them be more effective (if they see it that way)?
2. Are task level people in this area aware of the content of their leaders ' standard work?
3. Are leader standard work documents used as working "diaries" to record notes and observations? Do superiors meet with subordinate leaders to review these documents periodically? Ever? How often?
4. How often do this area's superiors review subordinate leaders' standard work for updating based on new issues and changes, e.g. resulting from accountability board tasks?
5. Is there a defined place where completed standard work documents are stored for a few months? Is it used?
6. Has leader standard work been used in this area to facilitate transitions between leaders?
7. Is leader standard work focused on compliance or improvement or balanced?

Assessment: Rate this area/areas from 1 to 5 on the scale below and note rationale for the rating

1. Pre-Lean	2. Starting	3. Recognizable	4. Stabilizing	5. Sustainable
No leader standard work (LSW) in place.	Leader standard work exists for a few positions. It's rarely carried, is followed sporadically. The original content has not been revised, refined. Most leaders view it as a check the box activity to drive compliance with defined processes w/ little or no emphasis on improvement.	Standard work exists for all line leaders in area: team, supervisor, mgr. Most have their standard work with them, follow it, use it as working record of the day. Most leaders can give examples illustrating how leader standard work has helped them and sustained improvements.	All leaders in the area carry, follow, and use their standard work as a daily working record. All superiors regularly review subordinate leaders' LSW documents with them weekly. All leaders can talk about how LSW benefits them and the process. LSW is revised to reflect and sustain process changes.	All transitions between leaders include review (possible revision), and walk through of LSW. All new leaders follow LSW from day one on job. Weekly LSW document review with superior used as monitoring, communication, and improvement method. Defined process for turn-in, storage of LSW documents.

Rationale for this rating:

Value Stream Mapping

Intent: Value Stream Maps (VSMs) should do two things:
- Show the step-by-step movement of information, patients, and or material through an area (or an entire value stream) that produces value for a customer, user, or patient - internal or external.
- Communicate process performance measures (safety, quality, time, cost), process problems, and improvement plans.

Diagnostic questions:
1. Are value stream maps visible here? If so, do they show current and planned future states and measures?
2. Are improvements planned for the area (or complete value streams) visibly posted? Can people explain them?
3. Are VSMs used to identify, communicate, track, and measure process improvements in the area?
4. Do VSMs show planned kaizens, completion status of kaizens, and improvement targets in current vs. future state performance measures? Can people explain the maps, kaizens, and measures?
5. Who prepares value stream maps here? How many of this area's leaders are proficient value stream mappers?

Assessment: Rate this area/areas from 1 to 5 on the scale below and note rationale for the rating

1. Pre-Lean	2. Starting	3. Recognizable	4. Stabilizing	5. Sustainable
No maps visible. Maps not used as part of area's improvement planning. Few if any in area know how to map.	Some tech specialists in area know how to map; most leaders do not. Maps, when present, show current state only. Maps may be posted but are out of date.	The area has visible plans for improvement; many of which shown on current and future state VSMs as planned or active kaizens. Some VSMs show current vs. future measures w/targets for improvement (such as in turnaround and throughput times, % value and time, patient safety incidents, productivity, uptime, yield, etc. Many people can explain the maps and measures. Many leaders are proficient mappers and draw their own VSMs.	Current state and 90-day future state maps showing improvement goals (measures) and activities (kaizens) are visible in the area. Most people can explain them. All leaders can map, use VSMs to systematically identify improvements large and small. Completion status of kaizens is shown on the VSMs, linked to project plans, and shown visually as status of progress against 90-day goals.	VSMs regularly used in the area's communications. Front line leaders teach VS mapping. All area leaders are proficient mappers. Area uses posted VSMs to show its improvement plans. The area's performance (down to the team) is reflected in the current state measures summary on its VSM (e.g. turnaround and throughput times, % value add time, safety and incidents, patient and customer satisfaction, productivity, uptime, yield.)

Rationale for this rating:

Process Definition

Intent: Process Definition should reflect two things:

• Line and support tasks should be documented and the documentation should be readily accessible.

• Documentation matches current practice; execution is consistent with documentation across people and shifts.

Diagnostic questions:

1. Are there documented definitions for all line and support processes? Where is the documentation located?

2. Is the documentation current; does it match actual practice?

3. Is standard work available for production tasks? For all levels of staffing, if applicable? Is it posted?

4. For repetitive processing areas, are operator balance charts available for each level of staffing, and posted in the areas they reflect?

5. Are definitions available, and posted, for tasks in the management process (e.g., who maintains tracking charts, standard meeting agendas, standard work for leaders, etc.)?

6. Are Job Instruction Training tools (job breakdown sheets) used for process documentation? For training? Who maintains them? Are they current? Examples?

Assessment: Rate this area/areas from 1 to 5 on the scale below and note rationale for the rating

1. Pre-Lean	2. Starting	3. Recognizable	4. Stabilizing	5. Sustainable
Process documentation either in binders or IT system not readily accessible. Most documentation is out of date - does not match actual practice.	Discussions in progress to update and convert documentation to useable format for a few areas on the floor. Some task/work balance charts visible, but most not current and for one staffing level. In repetitive areas, standard work w/ expected task times posted, but most out of date and/or for one takt pace.	Standard methods, procedures, step-by-step charts with expected times as applicable are visible in some areas for one level of staffing. In repetitive areas (e.g., processing or assembly), standard work or standardized procedure charts with times are available for some tasks/ work areas.	Most areas that operate w/multiple levels of staffing have task balance charts with expected times as applicable. Processes are defined for all production tasks and most regularly occurring management processes. Process documentation is kept at the point of use or application and is kept updated to match actual practice as improvements and changes occur.	Expected performance for all regularly occurring tasks and processes (even if infrequent) have been defined and documented. Process documentation is either displayed or accessible at point of use. Actual practice matches process documentation; evidence that documentation is updated to reflect changes in practice.

Rationale for this rating:

Process Discipline

Intent: Process Discipline should reflect two things:
- Line, support, and regularly occurring (even if infrequent) leadership tasks are documented.
- Actual practice reflects disciplined adherence to defined processes. Definitions are kept updated as processes change.

Diagnostic questions:

1. Are line, support, management processes defined? Regularly followed (e.g. training and qualification, repetitive production, changeovers/turnarounds, safety and housekeeping)?

2. Do crisis situations result in process shortcuts (e.g., material replenishment, qualified staff for defined tasks, changeover/turnarounds, holding areas for flow impediments)?

3. Are process assessments carried out? Regularly? How frequently? By those in the area or outsiders? Do internal as well as external assessment results produce improvements?

4. When assessments or cycle tracking turn up noncompliance or misses, are problem-solving tools used?

5. To what degree does process focus lead to process improvement and changes? Is there observable evidence?

6. How regularly do leaders conduct gemba walks to teach as well as to inspect? How many leaders do so?

Assessment: Rate this area/areas from 1 to 5 on the scale below and note rationale for the rating

1. Pre-Lean	2. Starting	3. Recognizable	4. Stabilizing	5. Sustainable
Leaders' attention is mostly focused on expectations for results. Consistent adherence to defined processes/expectations is almost totally lacking.	Processes are mostly followed when things run smoothly, but abandoned with high volume or when problems arise. A few leaders can speak to the Lean rationale for process discipline and sticking with it.	Most leaders focus on disciplined adherence in obvious processes such as frequently occurring or repetitive tasks and cycle tracking charts, a few also focus on discipline in lower volume/frequency and/or support processes. Most leaders do a good, clear, specific job of focusing on recorded process misses.	Leaders' focus (helped by cycle tracking charts) includes discipline in most line and support processes, including housekeeping, high and low volume production, changeovers/turnarounds, labor planning, material supply/replenishment. Most leaders using process tracking data to identify and act on improvement opportunities.	Regular and frequent reviews occur of production and support processes including regular process assessments to maintain adherence and identify improvement opportunities. All processes (line and support) track their performance and respond to misses with improvement task assignments and/or projects visible in the area.
Rationale for this rating:				

Process Improvement

Intent: Process Improvement should reflect two things:
- Everyone's job includes process improvement: line, support, admin people at all levels, floor to executive
- Improvement includes activities from small to large in scope, driven by process tracking and employee suggestions.

Diagnostic questions:
1. Who is usually involved in improvement: specialists, leaders, IT, support groups, suppliers, floor workers?
2. Who would most leaders say is responsible for process improvement?
3. How are assignments made for improvement tasks? Are the assignments and their status visually displayed?
4. How typical is it for improvement assignments to end up with actual improvements having been made?
5. Are kaizens a regular part of the improvement process in the area? Who participates; who leads them?
6. Does improvement work focus mostly on big, technically-led projects? Are small improvements pursued?
7. Is there a regular way for employees to suggest improvements? What % of employees make suggestions? How many are implemented: few, some, most, all?

Assessment: Rate this area/areas from 1 to 5 on the scale below and note rationale for the rating

1. Pre-Lean	2. Starting	3. Recognizable	4. Stabilizing	5. Sustainable
Improvements made by formal teams or in response to catastrophic failures. IT, Engineering, Finance, HR, other support groups lead improvement projects.	Project teams make small improvements during implementation debugging. Most (>75%) leaders see improvement as responsibility of technical support groups. Suggestion systems may be introduced but are not sustained.	Most leaders say they should be involved in process improvement; some actively support improvement throughout their areas. Many leaders use green/red daily accountability boards to drive improvement. Some tasks completed on time; some A3s used to track improvement projects. Most leaders have participated in kaizens, few have led, none facilitate kaizens.	Most leaders' clearly see process improvement within their responsibility, can give examples of their involvement. All leaders have been in kaizens, most now regularly lead kaizens. Most leaders effectively use daily or weekly task assignment boards, A3 project plan reviews as shown by audits of boards and completed tasks. Some leaders experimenting w/ employee suggestion systems.	Task assignments from regular stand up meetings regularly result in small and large improvements. Visual employee suggestion systems established, sustained w/steady flow of ideas, output of implemented improvements. Improvement plans, targets displayed on area info centers. Many leaders qualified kaizen facilitators. Lean resources team w/ rotating staffs support local improvement activities and Lean training.

Rationale for this rating:

Root Cause Problem Solving

Intent: Root Cause Problem Solving should reflect two things:
- "Problem solving" understood to mean eliminating source of a problem once and for all.
- When problems arise, leaders ask "Why?" and immediately or later initiate data-based root cause problem solving.

Diagnostic questions:
1. How often are workarounds used instead of investigating and resolving underlying causes of problems?
2. How often do leaders rely on data and analysis to attack a problem vs. gut feel, intuition, or impression?
3. To what degree do leaders expect changes will expose previously unseen problems that cannot be specifically anticipated, but proceed anyway?
4. How frequently do leaders ask why something happened vs. just asking what will we do to get back on track?
5. How frequently are leaders involved in leading problem-solving efforts?
6. How well and widely used are problem-solving tools such as five-whys, eight-step problem solving? Do leaders teach problem solving?
7. How frequently do leaders raise expectations for process performance in order to uncover the next level of process interruption or problem?

Assessment: Rate this area/areas from 1 to 5 on the scale below and note rationale for the rating

1. Pre-Lean	2. Starting	3. Recognizable	4. Stabilizing	5. Sustainable
Problem solving only focused on workarounds, not finding what caused the problem. Where cause analysis used, it is in formal technical project teams. Leaders can't describe problem solving, or if can, rarely if ever follow it.	Leaders have begun using visuals to collect problem data but w/little emphasis on cause analysis. Workarounds remain common response to problems. Evidence of one or few attempts at systematic problem solving. No leaders teach problem solving.	Some leaders beginning to ask why, pursue root causes for major problems, teach problem solving. Workarounds are recognized as such; evidence of problem-solving methods used to understand and attack causes. Uncovering flow interrupters still viewed as troubling surprises.	Many leaders asking why, pursuing root cause for big and small problems, beginning to use some form of structured problem solving - at least five Whys. Some teaching problem solving. Leaders expect changes to expose problems and to solve them at root cause level. Many leaders now seeking to improve their processes.	All leaders routinely expect cause analysis and pursuit of root causes for problems large and small. Most leaders teach problem solving. Process designs and measurements regularly tightened up to uncover the next level of problem: stated goal is to have perfect, zero waste processes.

Rationale for this rating:

Glossary

Balanced line: A series of workstations, such as an assembly line, where the time to do the work is nearly equal in each station within a few seconds. When a line is balanced to takt time, the time it takes to do the work in each station is equal to takt time or very close—within a second or two of takt. *See also* takt time.

Batch and queue: A method of organizing production. In batch and queue, the focus is on the efficiency of each discrete part of the production operation, such as a machine or a paint shop, as compared with the efficiency of the system as a whole from the first operation to final product. Each operation produces a batch of as many of a particular item at a time as possible, based on a separate production schedule sent to each operation. When the batch is done, it is pushed to the next operation regardless of the type and quantity of item needed by the next operation. The batch sits in queue, waiting for the next operation to get to it. Batch operations typically try to minimize the number of changeovers from one item to the next to maximize their efficiency in, for example, parts produced per machine per minute, hour, or day. *See also* changeover.

Brownfield: The name given to existing operations, particularly when they are in the process of being converted from batch-and-queue production to Lean production. Brownfield operations already have established practices and cultures, as distinct from greenfield operations, which often are literally constructed on what was previously an empty field. Greenfields have no history, no culture, and no preexisting practices.

Changeover: The time it takes to switch production of one product or component to making a different model, style, type of component, material, or finished product. Changeover time is often associated with machines or other equipment used to make or process components for several different product lines or styles. Changeovers consist of elements such as shutting down the machine or stopping production, removing the tool, die, or fixture for the first part, inserting the tool to make the second part, removing any remaining first parts and their component materials, bringing in the materials and containers to make and hold the second part, making a second part and checking it against specifications, and resuming production. In continuous process operations, changeovers can entail purging lines and equipment and changing process parameters before beginning the next run of material.

Countermeasure: In the Toyota Production System, when a problem arises that cannot be eliminated immediately, countermeasures are put in place to protect the process from the problem. When a countermeasure is in place, the problem is not considered fixed or solved; instead, it is considered just to be a Band-Aid® until a real solution is found that will eliminate the problem instead of simply covering it up. For example, holding inventory to protect against a machine that breaks down often is a countermeasure to be kept in place until the machine's reliability can be improved.

Culture: In a workplace, the sum of habits people rely on to get things done "the way we do things here." Culture also sums up the things an adult member of a work group needs to know and comply with in order not to be seen as deviant by other members. *See* Chapter 1.

Cycle time: *See* work content.

Discipline: What happens when someone breaks the rules or fails to meet expectations. The second meaning refers to adherence to defined processes, such as following the sequence of elements every time in standardized work, or filling in a pitch chart as each pitch cycle is completed. This latter meaning of discipline is especially crucial in a Lean system where processes are interdependent and closely tied with one another, and failure in any one can quickly bring the entire system to a halt.

FIFO: First in, first out. When units move between operations in FIFO order, they maintain their sequence. This can be important when work starts simultaneously, in different areas, on several subcomponents that will come together to form a single unit at the end of the production process. Maintaining FIFO sequence is also important in keeping stored inventory fresh by always using the oldest first.

Five S (5S): Sort, shine, set in order, standardize, sustain. These are five steps to remove unneeded tools, materials, debris, and clutter; thoroughly clean the entire area and everything in it; establish a logical place for each item; mark addresses/locations for each place and the thing(s) that go in it; and establish a system to maintain the cleanliness and order you have established. With 5S, everything has a place, and you can tell at a glance what is supposed to be where, what does not belong in an area, and what belongs, but is missing or out of place. It is one form of basic discipline.

Five whys (5Ws): Asking "why" five times is a basic method of cause analysis. Each successive question is intended to go deeper into the cause of a situation, typically a problem with production. For example: We missed pitch this period. Why? We ran out of parts. Why? The water spider didn't pick up the withdrawal card. Why? There was no withdrawal card in the container. Why? I took the card home in my pocket by accident and lost it. Why? I didn't put the withdrawal card in the kanban post when I pulled the container in to begin using it. What was the cause? Failure to handle the withdrawal kanban according to the defined process: put it in the kanban post before using the first part from the container.

Flow: A goal in Lean production in which product moves through the steps in a production process with no interruption or waiting between steps. An example of flow is a moving assembly line. The ideal of flow is one-piece flow. In one-piece flow, as work is completed at one workstation, the unit is passed directly to the next with no waiting and a new unit arrives at the first workstation without that operator having to wait for it. In some cases, flow is not one piece at a time, but rather a small batch or lot at a time, such as a pair of arms for a chair or a set of drawers for a storage cabinet.

Flow interrupters: Things that cause production to be halted or slowed below the expected pace, until they can be resolved. Running out of parts in an assembly process is a flow interrupter. So are machines breaking down unexpectedly, and unplanned absences of production team members.

Gemba walk: A Japanese term meaning "the real place," or roughly, "where the action is." In manufacturing, that means the production floor. In a customer service operation, gemba is where the agents are taking customer calls. In healthcare, gemba is the location of the area of focus such as the waiting room, registration stations, lab, pharmacy, or imaging, procedure, or surgical areas. Gemba walks are one of the primary ways for teaching Lean production, and the primary way for teaching Lean management. In a gemba walk, a teacher, or sensei, and student walk the production floor. The teacher asks the student to tell what he or she sees and, depending on the answer, asks more questions to stimulate the student to think differently about what is in front of him or her. This includes learning to see what is not there, and learning to see what has been accepted as a given, but should be changed to fit with Lean principles. Gemba walks often include assignments to act on what the student has come to see, with follow-up on the next gemba walk, typically weekly. *See also* sensei.

Heijunka: A Japanese term for a visual method used to smooth out demand for production so the demand is level, or stable every interval of time all day long. The intervals can be pitches, fractions of an hour, whole hours, or any other interval that fits a particular production process. In addition to leveling work across intervals, a heijunka system can also be used to introduce the same number and sequence of units of mixed types every interval, reflecting the overall proportions of mixed models to be produced in a given day or other production period. Heijunka leveling can be applied equally well to nonproduction activities, for example, when scheduling work to be carried out by a maintenance group, or controlling the work released into administrative or technical professional operations, for example, three low-work-content tasks per each high-content task, or responding to voicemail messages in the first ten minutes of each hour rather than letting them accumulate all day. *See also* mixed-model line.

Hypothetical construct: An idea or a label applied to an idea that cannot be directly observed. A construct is an abstraction such as market appeal, in distinction to something concrete, such as gross operating profit or first-year sales.

IT: An abbreviation for "information technology," meaning computers, computer automation, and computer networks. IT solutions to information flow and information management are often, but not always, at odds with the visual controls approach in Lean production and Lean management, where no computer is needed to assess the status of any process. Computer networks can be helpful in Lean systems by serving to broadcast or transmit information from one location to another, especially where line of sight communication is not possible or multiple configurations of options make simple visual methods less practical.

Job rotation: *See* rotation.

Kaizen: A Japanese term meaning "good change." Kaizen is a way of thinking and seeing, of always being alert to the opportunity to make changes for improvement. It typically involves small changes, rarely more than can be accomplished by a team of seven working full-time for a week. Many Lean organizations include kaizen events as a regular part of their continuous improvement activities. Kaizens are structured events carried out by a team assembled for the task under the direction of a kaizen team leader, where the improvement task is completed start to finish in a week or less.

Kanban: A Japanese term meaning "signboard." Kanbans are used to identify and order a specified quantity of parts. They include three kinds of information: the part or item number, the quantity, and the authorization either to make parts, in the case of a production instruction kanban, or to take parts, in the case of a withdrawal kanban. Kanbans are often printed cards, but can also be designated empty containers, or empty spots in a rack or on the floor. *See also* pull signals.

Layout: The arrangement of equipment and material storage in a production area, analogous to a floor plan for a room or house. Layouts in Lean production areas are designed to facilitate flow of material and production, with equipment arranged in the sequence of production steps. *See also* value stream.

Lean management system: The practices and tools used to monitor, measure, and sustain the operation of Lean production operations. Lean management practices identify where actual performance fails to meet expected performance, and assigns and follows up improvement activities to bring actual in line with expected, or to raise the level of expected performance. The basic components of the Lean management system are standard work for leaders, visual controls, and a daily accountability process.

Lean production: The name for the Toyota Production System popularized by Womack and Jones in *Lean Thinking*. Womack and Jones identify five elements, or principles, of Lean production: definition of value from the customer's point of view, flow production, pull replenishment of what has been consumed, constant emphasis on the reduction of waste, and striving for perfection.

Lot size: Lean practitioners often refer to lot sizes, rather than using the word *batch*. In any case, lot size refers to a specific quantity or batch in which parts are produced or procured. Batches, or lots of defined size, are used in Lean processes where one-piece flow is not yet possible because of bottlenecks or where resources are shared among value streams. The lot size (and container size) at the point where parts are used, typically in assembly, the lot size in which parts are held in supermarkets, and the lot sizes in which producing work centers make parts need to be in close alignment to avoid adding several potential forms of waste into the system as a whole. *See also* flow.

Machine balance chart: Shows the load on a machine as a stacked bar chart of the number of hours per week required to produce each type of part the machine makes, including the amount of time it is idle while changeovers from one part to the next are performed, and the amount of time it is typically unavailable because of breakdowns or planned maintenance. The chart reflects the time a machine must run and change over to produce specified quantities of the parts to replenish a supermarket without causing the supermarket to run out of any of the parts produced by the machine. A machine balance chart needs to be produced as part of the work involved in setting up a pull system's supermarket. *See also* pull system and supermarket.

Mixed-model line: An assembly or progressive build line on which several models of unit are produced, such as two different chair models, each of which can be ordered with different options, or different sizes of similar units like storage cabinets. So, instead of building a single finished good, the line is designed to produce several kinds of finished goods in a way that the mix does not interrupt flow. Heijunka scheduling is helpful for introducing the mix in a stable way that allows the line to maintain flow. *See* balanced line and heijunka.

MRP: Materials requirements planning (MRP) is the planning system for conventional batch-and-queue production systems. MRP forecasts the need for material and components based in part on historical patterns of demand and in part on current orders. MRP schedules work to be performed in batches to maximize efficiency of individual work centers. It is a classic push scheduling system.

Noncyclical processes: Processes not carried out every production cycle (such as every time a storage cabinet is assembled on an assembly line). Noncyclical tasks are those carried out one or only a few times a day, such as cleaning equipment at the end of a shift, or calibrating a process every 100 cycles. Noncyclical tasks can be even less frequent, for example, weekly, monthly, or quarterly maintenance tasks. Compare with standardized work.

Non-value adding: Tasks that do not transform parts or materials into finished products. Some non-value-adding work remains necessary in many production operations, such as parts traveling on a conveyor through a paint or finishing shop. Applying and curing the finish adds value, as it is part of the transformation of materials into products, but travel on the conveyor does not, even though it is necessary in the way the process is set up today. Tasks either add value or do not. People asked to do non-value-adding work are not themselves without value; they are only doing what the current system asks of them. Compare with value-adding activity.

OEE: Overall equipment effectiveness. OEE is a coefficient produced by multiplying three percentages: actual time a piece of equipment was available as a percentage of planned availability, actual yield of good output as a percentage of all output produced, and actual speed, or rate, of the equipment's operation as a percentage of its designed speed or rate. An OEE of 0.85 is considered excellent. OEE considers six categories of losses contributing to less than perfect (OE = 1.00) performance: waiting for operator, waiting for material, defective outputs, speed losses, minor stoppages, and major stoppages.

Pitch: A multiple of takt, first used as a unit of measure in the automotive supplier industry, where units such as door handles or mirrors are produced to order and shipped to assembly plants in standard containers, or packs. The pack quantity (for example, 24 mirrors) is designated as a pitch as a method to manage pace. If each mirror is produced to a takt time of 20 seconds, 24 units would equal 8 minutes of work. Instead of having to focus each 20 seconds, pitch allows a leader to check pace every 24 takt cycles in this example. Where standard pack quantities are not used, the concept is still useful. Pitch can be used to designate an interval of time at which to measure whether the actual pace of production equals what was expected. Pitch as a time interval can be an hour, half hour, quarter hour, or other interval depending on the nature of the product and maturity of the process. Where processing time varies, for example, from case to case in an interventional cardiology lab, an improvement-oriented focus would record delays and interruptions rather than expected versus actual time to complete a case. *See also* takt time.

Priority board: Part of a pull system for replenishing material that has been consumed by a customer process. Production instruction kanbans that direct a work center to make a specified quantity of a specific part are brought to the work center and placed on its priority board in the order that the cards arrive at the work center. The cards are, in effect, in line waiting their turn. As the work center operator completes a job, he or she takes the first card in line from the next-up position on the board and puts it in a designated spot (often marked "running") in the work center. The operator or team leader then moves each card up one spot on the board. *See* Figure 4.5 in Chapter 4 for an illustration.

Process focus: Virtually every process in a Lean system is defined, documented, and visually controlled. Each process in a Lean system is interdependent with one, several, or many other processes. Failure in any single process can quickly bring the system to a halt. Process focus is the practice and discipline of continually and regularly checking on the status of each process to be sure it is operating normally, or documenting when it is not, and identifying the cause of the abnormal condition and eliminating or preventing it from recurring. Process focus is a principal objective of the Lean management system.

Production instruction card: A type of kanban that authorizes the production of a specified quantity of a particular component or part. Production instructions are elements in pull production systems. They typically circulate between a supermarket and the work center where the part or component is produced when the level of that part in the supermarket reaches its reorder point. *See also* kanban and pull signals.

Progressive build line: Another name for assembly lines where a finished unit is built up piece by piece as it moves from one assembly station to the next. Progressive build does not require a moving conveyor line, only that a unit moves from station to station as it is assembled.

Pull, pull system: Used in Lean production systems where flow is not practical, pull production is based on replenishing what has been used by a customer, usually the next workstation in a production process, but sometimes an actual customer taking away finished goods. Pull production is typically used where a piece of equipment makes components for several product lines or value streams, often where the equipment produces components faster than a single consuming process can use them. Pull systems often involve supermarkets where components are kept in specified locations and quantities. As customers take components away from the market, signals to replenish what has been removed are sent to the producing work center. *See also* kanban and supermarket.

Pull signals, visual pull signals: Pull signals are the devices used to authorize and request the water spider to pick and deliver a specified quantity of a specific part, or to authorize and request a supplying work center to make a specified quantity of a specific part to replenish what has been taken away from the work center's supermarket. The pull signals are, classically, kanban cards, but can be empty racks or containers, empty spots on the floor or a designated location on a shelf, or space in a workstation. Pull signals can be fitted with bar codes to enable scanning and electronic ordering or recording, but for a pull system to remain consistent with a Lean philosophy, it must also remain visually verifiable. This means electronic pull signals (appropriate when the distance between supplier and customer is too great to confidently rely on the circulation of cards as the sole signal) and visual signals (the cards) can coexist, but visuals are always a requirement in Lean management. *See also* kanban and supermarket.

Queue: Material waiting to be processed is in queue, as in waiting its turn in line. Queues of material between operations are often referred to as WIP, or work in process inventory. Queues of WIP between processes are the inevitable result of producing components in batches as opposed to one at a time. *See also* batch and queue.

Root cause: The basic source from which a problem grows, as distinct from symptoms that are the visible effects of a problem. By doing a problem-solving analysis to find what is causing a problem, it is often possible to eliminate the cause altogether, or to prevent it from recurring. By analogy, if you cut the top off a weed, it is likely to grow back from the undisturbed root. If you dig the weed out by the root, it will not come back. *See also* five whys.

Rotation, rotation pattern: Refers to people moving from job to job at specified intervals or times during the day. Where some jobs involve higher levels of ergonomic stress, rotation might be as frequent as every half hour. In other situations, rotation might be less frequent, such as every hour, two hours, every break, at midshift, etc. Rotation serves three purposes: First, it mitigates the risk of repetitive motion injuries, letting people vary the muscle groups and postures from station to station. Second, it provides a broad base of skilled and experienced people able to step into any of several jobs when needed. Third, it provides many people's eyes on each and every job, improving the chances for thinking of and seeing potential process improvements.

Sensei: A Japanese term for one who has gone before, or teacher. The sensei is the master in the master-apprentice learning model for Lean management and for much of Lean production.

Setup: *See* changeover.

Standard work for leaders: One of the key elements of the Lean management system, standard work for leaders specifies the actions to be taken each day to focus on the processes in each leader's area of responsibility. Unlike standardized work for a production workstation, the elements in standard work for leaders usually are not timed, though some of them take place at specific times during the day, for example, "Lead tier two meeting (6:30 a.m.)." *See* Chapter 3 for a discussion of standard work for leaders.

Standardized work: Specifications, usually for a production workstation, that include the sequence in which steps or work elements are performed, expected time for each element and the total time for all, takt time, and the quantities of inventory before, in, and after the workstation. Standard work for production operations also can include specific safety information (for example, identifying potential pinch points) and quality checks to be performed during the sequence of operations.

Supermarket: Supermarkets are integral elements in Lean pull systems. They are areas where inventory is stored according to specific rules. Each item in the supermarket has a designated address as the only place where it is stored. Each item is stored in specific quantities, typically in a specified maximum quantity of containers. Each container is to hold a specified number of the designated item. The point at which an item is to be reordered, the reorder or trigger point, is clearly designated visually. In some cases, minimum quantities are also designated visually. One should be able to assess the status of a supermarket by using a simple visual inspection: Is the stock stored properly? Do part numbers correspond to the appropriate address? Are maximum quantity limits being adhered to? Are there dangerously low levels of some parts? Are the kanban cards in the proper locations according to the documented process? *See also* pull system and kanban.

Support groups: These are technical specialty departments found in most manufacturing plants. They can include engineering, production control (often called production and inventory control (PIC) or materials management), quality control or quality assurance (QC or QA), maintenance, tooling, safety, and human resources.

Takt time: The rate of customer demand calculated by dividing daily demand in units into the time available for production. It is typically expressed in seconds, as in "takt time for the XYZ is 54 seconds." That means a completed unit is due off the end of the line every 54 seconds. Lean systems are designed to produce at the takt rate; thus, takt is the basis for pace for the entire Lean process.

Team leader: Team leaders are the first level of formally designated leadership in a Lean production environment. Where production is done by an hourly workforce, team leaders are typically hourly employees who earn a slight premium for their position. The primary responsibilities of team leaders are to maintain the takt pace of production; maintain and improve on standardized work, including training operators in standard work; and be available for limited (5-minute) fill-in when a team member must leave his or her position in an emergency. Team leaders do not work production except in these situations. Team size is typically five to ten people working in a small area.

Technical Lean implementation: Focuses on changes to the layout of the shop floor, the way inventory is deployed and moves, and changes in how and where the production schedule is delivered to the shop floor. Typical areas of focus in technical implementations include establishing flow production balanced to standardized work at takt time, implementing pull systems with kanban signaling for replenishment, establishing water spider routes, and refining parts presentation and container size, among other things. Technical implementation differs from Lean management implementation.

Three-tier meetings: These make up the daily accountability process, one of the three key elements in the Lean management system. They are brief, structured, stand-up meetings that take place at or near the beginning of the workday. The first tier involves the work team led by the team leader. The second tier involves the team leaders led by the supervisor. The third tier involves the value stream staff led by the value stream manager. *See* Chapter 5 for a discussion of three-tier meetings.

Toyota Production System (TPS): Developed over the past 60 years at Toyota, based initially on the writings of Henry Ford. TPS seeks to eliminate waste from production processes. The ideal approach in TPS is for production to operate at exactly the rate of customer demand.

This is often expressed as just-in-time production, where nothing is produced until there is specific customer demand for it. *See also* pull, pull system, and flow.

Tugger: A small, usually electric engine often operated by a driver in a standing position. Tuggers are like small tram engines pulling carts or trailers of material around a production area. In Lean production systems, tuggers are often used to deliver materials from supermarkets to the workstations in assembly areas where the material will be used. This is especially true where the distances are too long to walk or the materials too heavy or large to transport by handcart.

Value-adding activity: Anything necessary to transform material on the way to making a finished product. Cutting to size, attaching components, making connections, and applying finish are all examples of value-adding activity. Storing, moving, counting, and reworking are all examples of activities that do not add value to the finished product, even though some of them may be necessary in the current production process. Compare with non-value adding.

Value stream: The people and equipment involved in producing a product line or closely related family of products. Value streams usually include each of the operations and pieces of equipment needed to make a product lined up or arranged close together in the sequence of production operations. The intent is to minimize the distance parts have to be moved and maximize the speed of flow through the production process. In a batch-and-queue organization, like operations are typically grouped together in separate departments, such as forming, molding, cutting, sewing, insertion, subassembly, finishing, and final assembly. Batches of parts typically wait, sometimes a long time, between operations and often must be moved long distances from one operation to the next.

Value stream manager: In an organizational structure designed to most fully reflect a Lean philosophy, all the support groups related to making a value stream operate would report on a solid line to the value stream manager. This means the value stream would have a dedicated staff group, as well as the typical line management positions, such as team leaders and department supervisors. Especially early in a Lean journey, companies may not be ready to make this organizational realignment without creating more turbulence than is worthwhile.

Visual controls: Visual controls are the variety of approaches that make the status of a process visible at a glance. They include production tracking charts of various kinds that show actual versus expected performance. They also include audit forms for status of safety practices, workplace organization, and 5S, compliance with specifications for contents of a supermarket, and number and location of kanban cards. Signs, labeled "parking spaces" on the floor itself, and shadow boards all make it possible to tell where things should and shouldn't be. Strictly speaking, visual controls allow control of processes rather than actually exerting control themselves. *See also* process focus.

Water spider: A name often used for the person who circulates between the points where material is consumed and the supermarkets where it is stored. Water spiders typically have defined routes and defined times, much like bus schedules. They pick up empty containers and kanban cards from the point of consumption, go to the supermarket and use the withdrawal kanban cards as a shopping list to pull full containers from the supermarket shelf or rack, and return to deliver the full containers. Water spiders sometimes remove finished goods from an assembly area, and sometimes take production instruction kanbans from the supermarket to the producing work center. *See also* supermarket and kanban.

Work breakdown structure: A basic component in project management, a work breakdown structure is essentially a list of all the steps or subtasks needed to accomplish a larger task. An example is a recipe where the steps are listed in sequence, one at a time, as opposed to simply listing the ingredients and stating, "Combine and bake."

Work content: Refers to the total amount of time required to perform all the elements in the standard work for a workstation. This is often referred to as the cycle time for a task. Total work content, also referred to as total cycle time, refers to the sum of cycle times of all workstations' standard work elements. Thus, the cycle times to complete the work elements in each of a series of nine workstations in a progressive build line might be 45 seconds in each workstation (though rarely are workstations perfectly balanced this way). The total cycle time would be 9 stations times 45 seconds for a total work content of 6 minutes 45 seconds.

Work element: *See* standardized work.

Bibliography

There are many references on Lean production. The ones I have listed here provide a good introduction to its concepts and application.

Dennis, Pascal. *Lean Production Simplified*. New York: Productivity Press, 2002.

Dillon, Andrew P. *The Sayings of Shigeo Shingo*. Cambridge, MA: Productivity Press, 1987.

Floyd, Raymond. *Liquid Lean*. New York: Productivity Press, 2010.

Imai, Masaaki. *Gemba Kaizen*. New York: McGraw-Hill, 1997.

Lean Enterprise Institute. *Lean Lexicon*. Brookline, MA: Lean Enterprise Institute, 2003.

Liker, Jeffrey K. *Becoming Lean*. New York: Productivity Press, 1998.

Rother, Mike, and Shook, John. *Learning to See*. Brookline, MA: Lean Enterprise Institute, 1998.

Schipper, Timothy, and Swets, Mark. *Innovative Lean Development*. New York: Productivity Press, 2009.

Shook, John. *Managing to Learn*. Cambridge, MA: Lean Enterprise Institute, 2008.

Womack, James, and Jones, Daniel. *Lean Thinking*, 2nd ed. New York: Simon and Schuster, 2003.

Bibliography

Index

About the Author

David Mann is the author of *Creating a Lean Culture: Tools to Sustain Lean Conversions*. The book was awarded the Shingo Prize for Operational Excellence in 2006 and has become a best seller in its field. It has been translated into Chinese, Polish, Portuguese, Russian, Spanish, and Thai. A second edition was published in 2010, and this third edition in 2015.

In 15 years of Lean experience at Steelcase, Inc., Mann developed and applied the concepts of a Lean management system supporting 40+ Lean manufacturing value stream transformations, and led an internal consulting team that supported over 100 successful Lean enterprise business process value stream conversions. He established a Lean consulting practice in 2005, and retired from Steelcase in 2009.

Mann's consulting, teaching, and coaching experience includes Lean transformation in manufacturing, enterprise business processes, and health-care organizations. His practice includes clients in healthcare, mining and energy, discrete and process manufacturing, technology, food processing, and enterprise business processes.

Mann is a frequent consultant trainer and speaker on Lean leadership and management, a Shingo Prize examiner, and a faculty member in management science at the Fisher College of Business, the Ohio State University.

Mann is an organizational psychologist, earning his Ph.D. at the University of Michigan in 1976. He lives in West Michigan with his wife, a retired criminal prosecutor. They have two daughters. For more information, visit www.dmannlean.com or contact him at dmann@dmannlean.com.

"David speaks with wisdom about the reality of the critical role of Leader Standard Work in process industry operations. He has hit the nail on the head regarding the challenge of creating leader time for Standard Work…an essential reference for any process industry leader tasked with managing the implementation of Lean within their organization…."

Kent Womack
Director, Operational Excellence, Mississippi Lime Company

"This third edition builds on the coherent description of a true Lean management system in the first two editions to help lean leaders understand how to engage executives in a way that embeds lean thinking throughout the organization. If you want your company to be great, I encourage you to study the ideas in this book."

Rebecca Morgan
President, Fulcrum ConsultingWorks, Inc., Cleveland, OH

"David demonstrates his commanding knowledge of Lean Culture through this book, where he breaks down a complex recipe into an easy to follow program that will lead organizations in any industry to a successful implementation."

Jesse Larrabee
Manager, Information Systems, Blue Hill Memorial Hospital

"Here is a resource every leader with a passion to apply scientific thinking to Lean should consider. It is written with his sensei-like clarity, which provides deep insights and illustrated approaches you can actually put to work. This is a book leaders will use and learn from. Included is access to Manufacturing and Healthcare standards, assessments, Gemba worksheets you can use immediately, and 52 case studies from manufacturing, healthcare, executives, and process industry sectors."

Dave Hogg
P. Eng., Editor, Canadian Manufacturers and Exporters